Handbook of Experimental Pharmacology

Volume 255

The *Handbook of Experimental Pharmacology* is one of the most authoritative and influential book series in pharmacology. It provides critical and comprehensive discussions of the most significant areas of pharmacological research, written by leading international authorities. Each volume in the series represents the most informative and contemporary account of its subject available, making it an unrivalled reference source.

HEP is indexed in PubMed and Scopus.

More information about this series at http://www.springer.com/series/164

Susan D. Brain • Pierangelo Geppetti
Editors

Calcitonin Gene-Related Peptide (CGRP) Mechanisms

Focus on Migraine

Editors
Susan D. Brain
School of Cardiovascular
Medicine & Sciences
King's College London
London, UK

Pierangelo Geppetti
Department of Health Sciences
University of Florence
Florence, Italy

ISSN 0171-2004   ISSN 1865-0325   (electronic)
Handbook of Experimental Pharmacology
ISBN 978-3-030-21456-2   ISBN 978-3-030-21454-8   (eBook)
https://doi.org/10.1007/978-3-030-21454-8

This Springer imprint is published by the registered company Springer Nature Switzerland AG.
The registered company address is: Gewerbestrasse 11, 6330 Cham, Switzerland

Preface

We have enjoyed creating this book at this timepoint where we are starting to see the therapeutic benefits of the many years of CGRP research, since its discovery nearly 40 years ago. Our aim has been to cover the major issues relating to CGRP in keeping with the long-standing focus of the *Handbook of Experimental Pharmacology* series. Our aim has also been to describe the development of knowledge concerning the action and potential of CGRP ligands, leading to the proof of its role in migraine and other headaches. There are some fascinating twists and turns in this story.

The first was the realization that this potent vasodilator peptide CGRP, discovered in the thyroid tissue of elderly rats, is primarily found in sensory nerves. Indeed, those very nerves had been investigated due to their links with another neuropeptide substance P. Early studies involved their comparison, but it became clear in time that CGRP had a very distinct profile, with CGRP being the most prominent in the pathophysiology of migraine. Not only was CGRP found in the trigeminovascular system but it was realized by Goadsby and Edvinsson that it is released into the jugular vein during migraine. Of importance its inhibition was observed with treatment with sumatriptan and as migraine resolved. Finally, the injection of CGRP can induce migraine.

The second concerned the discovery of the unique receptor for CGRP; this caused a few false starts, but the manuscript of McLatchie and co-workers still makes interesting reading. This volume has recent information on this intriguing receptor family that we still are only just beginning to understand. Finally, after the development of the first non-peptide antagonist by Doods and colleagues at Boehringer, it was realized that small molecule antagonists of the CGRP receptor benefitted migraine in a manner that does not implicate the worrisome vasoconstriction of the triptans. However, this positivity was soon dashed when it was realized that treatment with such small molecules was associated with a rise in the plasma levels of liver enzymes in certain of the patients.

At some point during this journey of discovery, it was realized that antibody therapy that was already in use for arthritis (e.g. the anti-TNF biologics) could be useful for migraine. There was intense debate in that it may be difficult for antibodies whose action is restricted to the periphery to target the migraine-provoking CGRP. Notwithstanding, several companies started to develop antibodies. Rumours started to surface in about 2012 that these antibodies were beneficial in migraine, with substantial evidence of benefit and negligible side effects by the time we started to plan this book. During its preparation we have seen their legal and ethical clearance for use in the treatment of migraine. There are now many reports of patients benefitting from this new class of therapy for migraine.

This has been a fantastic journey to date and we now look forward to the next stages and key questions, including the following: Will these antibodies be of use in other pain-related ailments? Can the prolonged use of such antibodies be considered as disease-modifying treatments? Will the orally available CGRP antagonists that remain in development 'push' the antibodies into a less prominent therapeutic place, due to their easier mode of administration?

We have assembled in this volume chapters from experts who have contributed to the development of our knowledge concerning CGRP. Many have been pivotal in the CGRP field. We thank everyone for their hard work in producing these manuscripts, which enables such a timely and complete volume. We would like to thank everyone, including those at Springer, who have worked so carefully to ensure the completed work for publication.

London, UK   Susan Brain
Florence, Italy   Pierangelo Geppetti

Contents

CGRP Discovery and Timeline

Kate Arkless, Fulye Argunhan, and Susan D. Brain

Contents

Abstract

Calcitonin gene-related peptide (CGRP) was discovered over about 35 years ago through molecular biological techniques. Its activity as a vasodilator and the proposal that it was involved in pain processing were then soon established. Today, we are in the interesting situation of having the approval for the clinical use of antagonists and antibodies that have proved to block CGRP activities and benefit migraine. Despite all, there is still much to learn concerning the relevance of the vasodilator and other activities as well as further potential applications of CGRP agonists and blockers in disease. This review aims to discuss the history and present knowledge and to act as an introductory chapter in this volume.

K. Arkless · F. Argunhan · S. D. Brain (✉)
Section of Vascular Biology & Inflammation, School of Cardiovascular Medicine & Sciences, BHF Centre for Cardiovascular Sciences, King's College London, London, UK
e-mail: Sue.brain@kcl.ac.uk

© Springer International Publishing AG, part of Springer Nature 2018   1
S. D. Brain, P. Geppetti (eds.), *Calcitonin Gene-Related Peptide (CGRP) Mechanisms*,
Handbook of Experimental Pharmacology 255, https://doi.org/10.1007/164_2018_129

Keywords
Cardiovascular · CGRP · Migraine · Review

1   Discovery of the Sensory Neuropeptide Calcitonin Gene-Related Peptide

Calcitonin gene-related peptide (CGRP) is a 37-amino acid peptide that was discovered in 1982 (Amara et al. 1982). It was realised that alternative processing of the RNA transcripts of the calcitonin gene leads to the generation of mRNA that encodes for CGRP instead of calcitonin. This discovery also revealed that calcitonin is the major product in healthy thyroid tissue, but a distinct mRNA precursor to CGRP predominates in the hypothalamus (Amara et al. 1982). This group, using recombinant DNA and molecular biology techniques, then quickly realised that CGRP is a neuropeptide. They also revealed its sensory neuronal localisation, in the rat trigeminal ganglia as well as sensory ganglia of the spine and in 'small beaded fibres' that innervated a range of organs and including the smooth muscle of blood vessels (Rosenfeld et al. 1983). They correctly suggested that it was involved in mediating functions that involved nociception, cardiovascular regulation and gastric-intestinal regulation (Rosenfeld et al. 1983). Indeed, they demonstrated that CGRP has potent effects on blood pressure regulation and catecholamine levels, primarily concentrating on the concept that CGRP acts on the central nervous system to stimulate sympathetic outflow of noradrenergic transmitters mediating increased blood pressure. They did, however, also show (but did not mention in the abstract) that CGRP caused a reduction in blood pressure. This was not further investigated except to point out that they observed a generalised vasodilation (Fisher et al. 1983). This manuscript emphasised their realisation that 'whilst this report is the first to describe biological actions of CGRP, future investigations are required to define its potential physiological roles as an intercellular transmitter'. This was swiftly followed by the publication of the structure of human CGRP, which had a high structural homology to rat CGRP (Morris et al. 1984) and its presence in the plasma of patients with medullary thyroid carcinoma (Morris et al. 1984). At the time, there was also confirmation that CGRP is localised to the spinal cord of humans and in a wide range of species (Gibson et al. 1984). It was then realised that human CGRP is a potent vasodilator, especially in the microcirculation (Brain et al. 1985). Indeed, it was suggested to be the most potent microvascular vasodilator known as the intradermal injection of femtomolar doses that induced a local erythema, due to increased blood flow that lasted for several hours in human skin (Brain et al. 1985). By comparison, substance P, prostaglandins and other microvascular vasodilators appeared substantially weaker (Brain et al. 1985). Later that year, it was shown that CGRP is a hypotensive agent in humans and associated with facial flushing in keeping with its potent microvascular effects (Gennari and Fischer 1985; Struthers et al. 1986). The examination of the effect of CGRP in human skin revealed that it was not only a more potent vasodilator than other known agents but also had a sustained duration of action (Brain et al. 1986b). Since the peripheral vasodilator activity of CGRP was discovered, there has been an understanding that CGRP can mediate vasodilation via endothelial-dependent and endothelial-independent mechanisms (Brain and Grant 2004). A range of signalling mechanisms by which CGRP may act in both vascular and non-vascular signalling have been reported (Russell et al. 2014) and are further examined in this book (Fig. 1).

Moreover, it was realised at an early stage that CGRP coexists with substance P in sensory nerves, which is of potential functional relevance, in terms of modulating the effects of inflammatory mediators and responses (Brain and Williams 1985; Lundberg et al. 1985). However, CGRP may also exist in nerves alone, especially Aδ fibres and in the CNS and has been suggested to be produced via several non-vascular cells in addition (Russell et al. 2014).

2 Family and Structure

It was soon discovered that the 37-amino acid peptide CGRP exists as two isoforms (α and β). These peptides are coded by separate regions of chromosome 11. They differ minimally in structure and species, possessing a disulphide bridge between the Cys2 and the Cys7 residues of the peptide. The β form is synthesised via a separate gene and originally considered to be found in the gut and brain (Steenbergh et al. 1986). These α and β forms (also known as I and II) have very similar activities, in terms of their vasodilator activity (Brain et al. 1986a; Franco-Cereceda et al. 1987). The CGRP peptides were realised, over the next few years, to be the members of a family that not only includes calcitonin but also the structurally related peptides amylin (AMY), adrenomedullin (AM) and intermedin, also called adrenomedullin-2 (AM2) and sometimes called adrenomedullin-2/intermedin (Russell et al. 2014). Whilst calcitonin plays an important role in calcium regulation, the other members of the family all possess cardiovascular properties as well as inflammatory/metabolic biological properties. All have more recently become the targets for potential therapeutic approaches and are heavily investigated. Of note, an amylin analogue (Pramlintide, marketed as Symilin in the USA) is administered by subcutaneous injection (Wysham et al. 2008). It is effective but has a short half-life.

3 Receptors

It took over a decade of intense research to discover the unique nature of the CGRP receptors. Whilst a truncated CGRP peptide ($CGRP_{8-37}$) was shown to have antagonist activity at an early stage in cardiovascular tissue, it did not have this activity in certain other tissues, leading to the suggestion that two receptors existed (Chiba et al. 1989). Independently, Kapas and co-workers suggested that two structurally related receptors RDC1 for CGRP and L1 for adrenomedullin existed (Kapas et al. 1995; Kapas and Clark 1995). However, these receptors could not be confirmed by other workers. Eventually in 1998, a key publication revealed that CGRP acts through a receptor complex (McLatchie et al. 1998). The main subunit of the complex is a

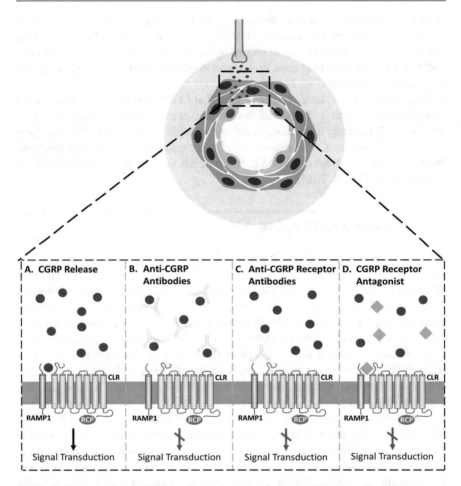

Fig. 1 Possible mechanisms to block calcitonin gene-related peptide (CGRP) signalling. (**a**) The neuropeptide, CGRP, is released from sensory nerves and acts through a heterodimeric receptor complex, composed of the calcitonin receptor-like receptor (CLR) and receptor activity-modifying protein (RAMP1), along with an intracellular receptor component protein (RCP). Upon binding, signal transduction occurs, leading to vasodilation and other cardioprotective mechanisms. (**b**) One possible method of blocking CGRP is the use of anti-CGRP antibodies, such as LY2951742. Here, free CGRP is no longer available to bind its receptor complex, thus blocking signal transduction. In this scenario, however, other molecules with affinity for the CGRP receptor complex, such as adrenomedullin, may bind instead. (**c**) Antibodies can also be raised against the CGRP receptor complex to block signalling. For example, the human monoclonal antibody, AMG 334 (Erenumab) is currently in Phase III clinical trials for treatment of migraine. Blocking of this complex may cause CGRP to bind other receptors for which it has affinity, such as the amylin 1 receptor complex. (**d**) CGRP receptor antagonists, such as the small molecule inhibitor, BIBN4096BBS may also be used to block CGRP signalling but, again, any remaining free CGRP may still be able to bind other receptors

G-protein calcitonin receptor-like receptor (CLR) of the family B of G-protein coupled receptors. However, CLR could only act when linked to a single transmembrane protein that became known as receptor activity-modifying protein (RAMP). It was realised from cell studies that there are three RAMPs (RAMP1, RAMP2 and

RAMP3) and that the co-localisation at the cell membrane of RAMP1 with CLR revealed a functional CGRP receptor (McLatchie et al. 1998). Moreover, it soon became established that CLR and RAMP1 comprise a receptor with high affinity for CGRP, CLR and RAMP2 comprise a receptor with high affinity for adrenomedullin and CLR and RAMP3 comprise a receptor with various agonists of the CGRP family (McLatchie et al. 1998). Unlike CGRP and adrenomedullin, intermedin does not appear to preferentially stimulate a receptor and is described as a non-selective agonist of the CGRP receptors (Roh et al. 2004). The RAMP proteins are important for aiding the movement of CLR to the cell membrane and through interaction with CLR provide a binding site for ligands (McLatchie et al. 1998). This is a somewhat unique system allowing different pharmacology to be exhibited by each receptor, which will be further discussed in this book. The receptor is further complicated by the presence of a receptor component protein (RCP). RCP enhances the efficiency of the signalling of the CGRP receptor (Evans et al. 2000), most probably via Gs, although a range of other pathways exist (Walker et al. 2010).

One highly significant finding was that of the realisation that the amino acid residue 74, a basic tryptophan in non-primates, is a tryptophan in humans, leading to species selectively with respect to antagonists (Mallee et al. 2002). Meanwhile, amylin associates with the calcitonin receptor CTR and RAMPs, thus providing a family of receptors that are less well understood and will be discussed further in this book. Intriguingly, CGRP has a similar affinity to one of the amylin receptors (AMY1(a), comprised of the CTR and RAMP1 as for the CGRP receptor); the in vivo relevance is unclear, but further discussed (Walker et al. 2015; Hay et al. 2018).

4   Migraine

An early indicator that CGRP could act in the cerebral circulation was given by Hanko and co-workers studying the pial vessels of animals and humans in vitro (Hanko et al. 1985). It was found to be present in the trigeminal ganglion in rat, in more neurons than substance P (Lee et al. 1985). Edvinsson and colleagues revealed the ability of CGRP to relax cerebral vessels via cAMP as well as via other endothelial-dependent mechanisms (Edvinsson et al. 1985) and that it was more potent than substance P, indicating its potential physiological significance in the trigeminocerebral circulation, although migraine was not mentioned (McCulloch et al. 1986). The possibility that CGRP was involved in extracranial pain was raised (Uddman et al. 1986). A key step forward was the realisation that CGRP could be released into the jugular vein of cats after electrical stimulation of the trigeminovascular system, a model associated with migraine and in humans, during an operation involving thermo-coagulation stimulation, where facial flushing was also observed in some individuals (Goadsby et al. 1988). The use of a novel intravital microscopy rodent model involving a closed cranial window revealed that CGRP, rather than substance P, was responsible for the vasodilation (Williamson et al. 1997). The CGRP-induced dural vasodilation was suggested to mediate trigeminal sensitisation that was reduced in the presence of a 5HT1B/1D agonist (Cumberbatch et al. 1999); thus facilitating development of the hypothesis that migraine is a neurovascular disease.

It was shown using similar techniques in 1993 that in the animal model, the stimulation of the trigeminal ganglion was associated with increased blood flow and release of CGRP that was inhibited when either sumatriptan or dihydroergotamine were administered (Goadsby and Edvinsson 1993). A link between the therapeutic benefit and the effect of these drugs on reducing the release and action of CGRP was important. The human study revealed that sumatriptan, in addition to relieving the migraine, was highly associated with reduced levels of CGRP measured in the jugular vein (Goadsby and Edvinsson 1993).

CGRP has also been shown to induce migraine-like symptoms after intravenous infusion (Lassen et al. 2002) and this is further discussed in the book. This was perhaps some of the first key evidence that CGRP may act peripherally to affect the cerebral circulation in terms of enhancing pain. However, by comparison CGRP injected into the human forearm stimulates a sustained local increased blood flow, but not nociception in terms of either pain or itch (Brain et al. 1986b; Pedersen-Bjergaard et al. 1991).

5   Antagonists

The first non-peptide CGRP antagonist Olcegepant (BIBN4096BS) was developed by Doods and co-workers. It exhibited an affinity for CGRP receptors in the pM range and inhibited the facial flushing after stimulation of the trigeminal ganglion, in primates (Doods et al. 2000), as previously shown with CGRP. It was found to be effective as an anti-migraine treatment in humans for up to 6 h after onset (Olesen et al. 2004). This book will further chart the development of antagonists and antibodies since then.

Olcegepant has been described as the first potential anti-migraine therapy that is not a vasoconstrictor and indeed this has been supported by more recent studies. Telcagepant (MK-0974) which was available orally confirmed this promise. However, hepatic adverse indications were revealed through elevated plasma transaminase during phase III clinical trials (Ho et al. 2016). There remains a question over whether this adverse effect is due to the effects of inhibiting endogenous CGRP, or related to the chemistry of this specific type of CGRP antagonist (Gottschalk 2016). Importantly, supporting the latter suggestion, recently ubrogepant (MK-1602) has been shown to be effective in the acute treatment of migraine, without effects on the liver (Voss et al. 2016). The overall high tolerability of the CGRP antagonists, apart from the hepatic adverse events, is good and now well documented (Bigal et al. 2013; Gottschalk 2016). Furthermore, there is good evidence that CGRP antagonists do not affect blood pressure, under circumstances examined in humans to date.

6   Antibodies

Clinical trials for monoclonal antibodies with CGRP have been equally successful, when compared with the antagonists in benefitting migraine, and have been substantially reviewed elsewhere (Bigal et al. 2013; Deen et al. 2017). Positive clinical trials have been published for four antibodies to date. These are Galcanezumab, LY2951742 (Lilly), Eptinezumab, ALD403 (Alder) and Fremanezumab, LBR-101 (Teva), which all target the peptide and finally Erenumab AMG334 (Amgen), which targets the CGRP receptor. The positive results with AMG334, in a phase 2 trial, are indicative that the CGRP receptor plays a primary role in migraine (Sun et al. 2016). These compounds may both prevent and benefit migraine. To date, there have been no hepatic adverse effects observed. These will be further discussed in this book. The trials with some of these compounds have now progressed to a stage where we have approval for some of these compounds for clinical use.

There is substantial excitement concerning the antibodies as the clinical trials have shown benefit in several types of migraine, with a low incidence of mild adverse effects. It is unlikely that antibodies cross the blood–brain barrier, thus it is widely thought that agents act via peripheral mechanisms. This understanding has led to a range of hypotheses being developed for the site of action that will be further explored in this book.

7   Other Indications for Calcitonin Gene-Related Peptide Antagonists and Antibodies

It has proved difficult to fully understand the physiological roles of CGRP, due to its broad localisation that encompasses practically all tissues of the body. It is therefore not surprising that it has been suggested to play a large role in a number of syndromes [recent review is given by Russell et al. (2014)] and in the neuro-immune axis (Assas et al. 2014). CGRP has also been suggested to play a central role in skin blushing and in various situations associated with menopause (Sharma et al. 2010; Hay and Poyner 2009; Russell et al. 2014). Potentially, CGRP antagonists may be of use in these syndromes. CGRP is suggested to be involved in painful arthritis from rodent studies (Nieto et al. 2015; Bullock et al. 2014); however, there is no evidence of a beneficial role of CGRP antagonists/antibodies to date. A clinical trial was carried out involving patients with osteo-arthritis knee pain, and the monoclonal antibody LY2951742, but unlike treatment with the positive control celecoxib, LY2951742 did not provide benefits under the conditions examined (Jin et al. 2016). Further information is given in a systematic review covering the subject of 'CGRP and non-headache pain' (Schou et al. 2017) and in this book.

8   Role of Calcitonin Gene-Related Peptide in the Cardiovascular System

The development of CGRP antagonists and antibodies that are beneficial in clinical trials involving migraines allows us to consider what other primary roles CGRP may have. Perhaps surprisingly, neither CGRP antagonists nor antibodies have significant effects on the physiological regulation of blood pressure or in influencing tendencies to peripheral vasoconstriction [reviewed in Bigal et al. (2013) and Russell et al. (2014)]. This indicates that CGRP is not a functionally important regulator of vascular tone. This would seem surprising considering its wide distribution within the cardiovascular system, potent vasodilator activity and evident presence of CGRP receptors. One possibility is that CGRP is only protective when the cardiovascular system is stressed and this is the subject of ongoing debate (MaassenVanDenBrink et al. 2016). This is possible as clinical studies to date have involved humans those in the majority have a healthy cardiovascular system. However, the studies carried out to date include exercising on a treadmill with patients suffering from angina (Chaitman et al. 2012). These authors suggested that CGRP was redundant in their model, where other endogenous vasodilators presumably take a more vital role (Chaitman et al. 2012). The concept that wiping out CGRP may be detrimental to cardiovascular function is further explored in this book.

9   The Therapeutic Potential of Calcitonin Gene-Related Peptide Agonists

The previous sections outline scenarios that lead to the possibility that there may be insufficient endogenous CGRP, released in humans for its blockade, to markedly affect the cardiovascular system. This may perhaps, in part due to evolutionary progress. Alternatively, it has been shown that endogenous CGRP levels can decrease with disease, for example, plasma levels in chronic congestive heart failure (Taquet et al. 1992). This may be due to increased sensory activity during the disease subsequently leading to a loss of sensory nerve function due to depletion/desensitisation. This is despite receptors for CGRP being present in the vasculature in an active form and able to respond to endogenous ligands. If this is the case, then CGRP agonists may potentially have a beneficial role in protecting against the harm and severity of cardiovascular diseases such as heart failure. Indeed, there is some historical evidence that this may occur in humans. Firstly, Gennari and colleagues gave intravenous CGRP to five patients with heart failure. Whilst there were small effects on blood pressure and heart rate, an improved contractility was observed in all (Gennari et al. 1990). In a separate study, Shekhar and colleague studied an 8-h infusion time in nine patients with congestive heart failure, without evidence of tolerance. CGRP was only effective whilst being infused and acted to increase cardiac output, stroke volume and, interestingly, renal blood flow. However, there was little effect on blood pressure (Shekhar et al. 1991).

Finally, CGRP can dampen the immune responses in a variety of situations (Russell et al. 2014; Assas et al. 2014); but it is difficult to determine where and whether these effects may be pivotal in humans. Three interesting possibilities that are under current investigation are that firstly, CGRP has been shown to reduce Langerhans cell-induced HIV-1 transmission and to enhance proteasomal degradation (Ganor et al. 2013; Bomsel and Ganor 2017); secondly, CGRP has been suggested to mediate immuno-suppression in lung infections (Baral et al. 2018); and thirdly, CGRP may play a primary sensing role, that is protective and includes the parabrachial nerves in the CNS (Campos et al. 2018).

10 Conclusion

This introductory chapter acts as a review of current knowledge concerning CGRP in physiology and disease and the results from clinical trials using CGRP antagonists and antibodies to date. The potential of these concepts will be further discussed in this volume.

Acknowledgements We thank the UK Medical Research Council Doctoral Training Partnership for funding for MRes/PhD studentships scheme by which Fulye Argunhan and Kate Arkless are supported.

References

Amara SG, Jonas V, Rosenfeld MG, Ong ES, Evans RM (1982) Alternative RNA processing in calcitonin gene expression generates mRNAs encoding different polypeptide products. Nature 298:240–244

Assas BM, Pennock JI, Miyan JA (2014) Calcitonin gene-related peptide is a key neurotransmitter in the neuro-immune axis. Front Neurosci 8:23. https://doi.org/10.3389/fnins.2014.00023

Baral P, Umans BD, Li L, Wallrapp A, Bist M, Kirschbaum T, Wei Y, Zhou Y, Kuchroo VK, Burkett PR, Yipp BG, Liberles SD, Chiu IM (2018) Nociceptor sensory neurons suppress neutrophil and gammadelta T cell responses in bacterial lung infections and lethal pneumonia. Nat Med 24:417–426. https://doi.org/10.1038/nm.4501

Bigal ME, Walter S, Rapoport AM (2013) Calcitonin gene-related peptide (CGRP) and migraine current understanding and state of development. Headache 53:1230–1244. https://doi.org/10.1111/head.12179

Bomsel M, Ganor Y (2017) Calcitonin gene-related peptide induces HIV-1 proteasomal degradation in mucosal Langerhans cells. J Virol 91. https://doi.org/10.1128/JVI.01205-17

Brain SD, Grant AD (2004) Vascular actions of calcitonin gene-related peptide and adrenomedullin. Physiol Rev 84:903–934. https://doi.org/10.1152/physrev.00037.2003

Brain SD, Williams TJ (1985) Inflammatory oedema induced by synergism between calcitonin gene-related peptide (CGRP) and mediators of increased vascular permeability. Br J Pharmacol 86:855–860

Brain SD, Williams TJ, Tippins JR, Morris HR, Macintyre I (1985) Calcitonin gene-related peptide is a potent vasodilator. Nature 313:54–56

Brain SD, Macintyre I, Williams TJ (1986a) A second form of human calcitonin gene-related peptide which is a potent vasodilator. Eur J Pharmacol 124:349–352

Brain SD, Tippins JR, Morris HR, Macintyre I, Williams TJ (1986b) Potent vasodilator activity of calcitonin gene-related peptide in human skin. J Invest Dermatol 87:533–536

Bullock CM, Wookey P, Bennett A, Mobasheri A, Dickerson I, Kelly S (2014) Peripheral calcitonin gene-related peptide receptor activation and mechanical sensitization of the joint in rat models of osteoarthritis pain. Arthritis Rheumatol 66:2188–2200. https://doi.org/10.1002/art.38656

Campos CA, Bowen AJ, Roman CW, Palmiter RD (2018) Encoding of danger by parabrachial CGRP neurons. Nature 555:617–622. https://doi.org/10.1038/nature25511

Chaitman BR, Ho AP, Behm MO, Rowe JF, Palcza JS, Laethem T, Heirman I, Panebianco DL, Kobalava Z, Martsevich SY, Free AL, Bittar N, Chrysant SG, Ho TW, Chodakewitz JA, Murphy MG, Blanchard RL (2012) A randomized, placebo-controlled study of the effects of telcagepant on exercise time in patients with stable angina. Clin Pharmacol Ther 91:459–466. https://doi.org/10.1038/clpt.2011.246

Chiba T, Yamaguchi A, Yamatani T, Nakamura A, Morishita T, Inui T, Fukase M, Noda T, Fujita T (1989) Calcitonin gene-related peptide receptor antagonist human CGRP-(8-37). Am J Physiol 256:E331–E335. https://doi.org/10.1152/ajpendo.1989.256.2.E331

Cumberbatch MJ, Williamson DJ, Mason GS, Hill RG, Hargreaves RJ (1999) Dural vasodilation causes a sensitization of rat caudal trigeminal neurones in vivo that is blocked by a 5-HT1B/1D agonist. Br J Pharmacol 126:1478–1486. https://doi.org/10.1038/sj.bjp.0702444

Deen M, Correnti E, Kamm K, Kelderman T, Papetti L, Rubio-Beltran E, Vigneri S, Edvinsson L, Maassen Van Den Brink A, European Headache Federation School of Advanced, S (2017) Blocking CGRP in migraine patients – a review of pros and cons. J Headache Pain 18:96. https://doi.org/10.1186/s10194-017-0807-1

Doods H, Hallermayer G, Wu D, Entzeroth M, Rudolf K, Engel W, Eberlein W (2000) Pharmacological profile of BIBN4096BS, the first selective small molecule CGRP antagonist. Br J Pharmacol 129(3):420. https://doi.org/10.1038/sj.bjp.0703110

Edvinsson L, Fredholm BB, Hamel E, Jansen I, Verrecchia C (1985) Perivascular peptides relax cerebral arteries concomitant with stimulation of cyclic adenosine monophosphate accumulation or release of an endothelium-derived relaxing factor in the cat. Neurosci Lett 58:213–217

Evans BN, Rosenblatt MI, Mnayer LO, Oliver KR, Dickerson IM (2000) CGRP-RCP, a novel protein required for signal transduction at calcitonin gene-related peptide and adrenomedullin receptors. J Biol Chem 275:31438–31443. https://doi.org/10.1074/jbc.M005604200

Fisher LA, Kikkawa DO, Rivier JE, Amara SG, Evans RM, Rosenfeld MG, Vale WW, Brown MR (1983) Stimulation of noradrenergic sympathetic outflow by calcitonin gene-related peptide. Nature 305:534–536

Franco-Cereceda A, Gennari C, Nami R, Agnusdei D, Pernow J, Lundberg JM, Fischer JA (1987) Cardiovascular effects of calcitonin gene-related peptides I and II in man. Circ Res 60:393–397

Ganor Y, Drillet-Dangeard AS, Lopalco L, Tudor D, Tambussi G, Delongchamps NB, Zerbib M, Bomsel M (2013) Calcitonin gene-related peptide inhibits Langerhans cell-mediated HIV-1 transmission. J Exp Med 210:2161–2170. https://doi.org/10.1084/jem.20122349

Gennari C, Fischer JA (1985) Cardiovascular action of calcitonin gene-related peptide in humans. Calcif Tissue Int 37:581–584

Gennari C, Nami R, Agnusdei D, Fischer JA (1990) Improved cardiac performance with human calcitonin gene related peptide in patients with congestive heart failure. Cardiovasc Res 24:239–241

Gibson SJ, Polak JM, Bloom SR, Sabate IM, Mulderry PM, Ghatei MA, Mcgregor GP, Morrison JF, Kelly JS, Evans RM et al (1984) Calcitonin gene-related peptide immunoreactivity in the spinal cord of man and of eight other species. J Neurosci 4:3101–3111

Goadsby PJ, Edvinsson L (1993) Examination of the involvement of neuropeptide Y (NPY) in cerebral autoregulation using the novel NPY antagonist PP56. Neuropeptides 24:27–33

Goadsby PJ, Edvinsson L, Ekman R (1988) Release of vasoactive peptides in the extracerebral circulation of humans and the cat during activation of the trigeminovascular system. Ann Neurol 23:193–196. https://doi.org/10.1002/ana.410230214

Gottschalk PC (2016) Telcagepant – almost gone, but not to be forgotten (invited editorial related to Ho et al., 2015). Cephalalgia 36:103–105. https://doi.org/10.1177/0333102415584311

Hanko J, Hardebo JE, Kahrstrom J, Owman C, Sundler F (1985) Calcitonin gene-related peptide is present in mammalian cerebrovascular nerve fibres and dilates pial and peripheral arteries. Neurosci Lett 57:91–95

Hay DL, Poyner DR (2009) Calcitonin gene-related peptide, adrenomedullin and flushing. Maturitas 64:104–108. https://doi.org/10.1016/j.maturitas.2009.08.011

Hay DL, Garelja ML, Poyner DR, Walker CS (2018) Update on the pharmacology of calcitonin/ CGRP family of peptides: IUPHAR review 25. Br J Pharmacol 175:3–17. https://doi.org/10.1111/bph.14075

Ho TW, Ho AP, Ge YJ, Assaid C, Gottwald R, Macgregor EA, Mannix LK, Van Oosterhout WP, Koppenhaver J, Lines C, Ferrari MD, Michelson D (2016) Randomized controlled trial of the

CGRP receptor antagonist telcagepant for prevention of headache in women with perimenstrual migraine. Cephalalgia 36:148–161. https://doi.org/10.1177/0333102415584308

Jin Y, Smith C, Monteith D, Brown R, Camporeale A, Mcnearney T, Deeg M, Raddad E, De La Pena A, Kivitz A, Schnitzer T (2016) Ly2951742, a monoclonal antibody against CGRP, failed to reduce signs and symptoms of knee osteoarthritis. Osteoarthr Cartil 24:S50–S50. https://doi.org/10.1016/j.joca.2016.01.114

Kapas S, Clark AJ (1995) Identification of an orphan receptor gene as a type 1 calcitonin gene-related peptide receptor. Biochem Biophys Res Commun 217:832–838. https://doi.org/10.1006/bbrc.1995.2847

Kapas S, Catt KJ, Clark AJ (1995) Cloning and expression of cDNA encoding a rat adrenomedullin receptor. J Biol Chem 270:25344–25347

Lassen LH, Haderslev PA, Jacobsen VB, Iversen HK, Sperling B, Olesen J (2002) CGRP may play a causative role in migraine. Cephalalgia 22:54–61. https://doi.org/10.1046/j.1468-2982.2002.00310.x

Lee Y, Kawai Y, Shiosaka S, Takami K, Kiyama H, Hillyard CJ, Girgis S, Macintyre I, Emson PC, Tohyama M (1985) Coexistence of calcitonin gene-related peptide and substance P-like peptide in single cells of the trigeminal ganglion of the rat: immunohistochemical analysis. Brain Res 330:194–196

Lundberg JM, Franco-Cereceda A, Hua X, Hokfelt T, Fischer JA (1985) Co-existence of substance P and calcitonin gene-related peptide-like immunoreactivities in sensory nerves in relation to cardiovascular and bronchoconstrictor effects of capsaicin. Eur J Pharmacol 108:315–319

Maassenvandenbrink A, Meijer J, Villalon CM, Ferrari MD (2016) Wiping out CGRP: potential cardiovascular risks. Trends Pharmacol Sci 37:779–788. https://doi.org/10.1016/j.tips.2016.06.002

Mallee JJ, Salvatore CA, Lebourdelles B, Oliver KR, Longmore J, Koblan KS, Kane SA (2002) Receptor activity-modifying protein 1 determines the species selectivity of non-peptide CGRP receptor antagonists. J Biol Chem 277:14294–14298. https://doi.org/10.1074/jbc.M109661200

Mcculloch J, Uddman R, Kingman TA, Edvinsson L (1986) Calcitonin gene-related peptide: functional role in cerebrovascular regulation. Proc Natl Acad Sci U S A 83:5731–5735

Mclatchie LM, Fraser NJ, Main MJ, Wise A, Brown J, Thompson N, Solari R, Lee MG, Foord SM (1998) RAMPs regulate the transport and ligand specificity of the calcitonin-receptor-like receptor. Nature 393:333–339. https://doi.org/10.1038/30666

Morris HR, Panico M, Etienne T, Tippins J, Girgis SI, Macintyre I (1984) Isolation and characterization of human calcitonin gene-related peptide. Nature 308:746–748

Nieto FR, Clark AK, Grist J, Chapman V, Malcangio M (2015) Calcitonin gene-related peptide-expressing sensory neurons and spinal microglial reactivity contribute to pain states in collagen-induced arthritis. Arthritis Rheumatol 67:1668–1677. https://doi.org/10.1002/art.39082

Olesen J, Diener HC, Husstedt IW, Goadsby PJ, Hall D, Meier U, Pollentier S, Lesko LM, Group, BBCPOCS (2004) Calcitonin gene-related peptide receptor antagonist BIBN 4096 BS for the acute treatment of migraine. N Engl J Med 350:1104–1110. https://doi.org/10.1056/NEJMoa030505

Pedersen-Bjergaard U, Nielsen LB, Jensen K, Edvinsson L, Jansen I, Olesen J (1991) Calcitonin gene-related peptide, neurokinin A and substance P: effects on nociception and neurogenic inflammation in human skin and temporal muscle. Peptides 12:333–337

Roh J, Chang CL, Bhalla A, Klein C, Hsu SY (2004) Intermedin is a calcitonin/calcitonin gene-related peptide family peptide acting through the calcitonin receptor-like receptor/receptor activity-modifying protein receptor complexes. J Biol Chem 279:7264–7274. https://doi.org/10.1074/jbc.M305332200

Rosenfeld MG, Mermod JJ, Amara SG, Swanson LW, Sawchenko PE, Rivier J, Vale WW, Evans RM (1983) Production of a novel neuropeptide encoded by the calcitonin gene via tissue-specific RNA processing. Nature 304:129–135

Russell FA, King R, Smillie SJ, Kodji X, Brain SD (2014) Calcitonin gene-related peptide: physiology and pathophysiology. Physiol Rev 94:1099–1142. https://doi.org/10.1152/physrev.00034.2013

Schou WS, Ashina S, Amin FM, Goadsby PJ, Ashina M (2017) Calcitonin gene-related peptide and pain: a systematic review. J Headache Pain 18:34. https://doi.org/10.1186/s10194-017-0741-2

Sharma S, Mahajan A, Tandon VR (2010) Calcitonin gene-related peptide and menopause. J Midlife Health 1:5–8. https://doi.org/10.4103/0976-7800.66985

Shekhar YC, Anand IS, Sarma R, Ferrari R, Wahi PL, Poole-Wilson PA (1991) Effects of prolonged infusion of human alpha calcitonin gene-related peptide on hemodynamics, renal blood flow and hormone levels in congestive heart failure. Am J Cardiol 67:732–736

Steenbergh PH, Hoppener JW, Zandberg J, Visser A, Lips CJ, Jansz HS (1986) Structure and expression of the human calcitonin/CGRP genes. FEBS Lett 209:97–103

Struthers AD, Brown MJ, Macdonald DW, Beacham JL, Stevenson JC, Morris HR, Macintyre I (1986) Human calcitonin gene related peptide: a potent endogenous vasodilator in man. Clin Sci (Lond) 70:389–393

Sun H, Dodick DW, Silberstein S, Goadsby PJ, Reuter U, Ashina M, Saper J, Cady R, Chon Y, Dietrich J, Lenz R (2016) Safety and efficacy of AMG 334 for prevention of episodic migraine: a randomised, double-blind, placebo-controlled, phase 2 trial. Lancet Neurol 15:382–390. https://doi.org/10.1016/S1474-4422(16)00019-3

Taquet H, Komajda M, Grenier O, Belas F, Landault C, Carayon A, Lechat P, Grosgogeat Y, Legrand JC (1992) Plasma calcitonin gene-related peptide decreases in chronic congestive heart failure. Eur Heart J 13:1473–1476

Uddman R, Edvinsson L, Jansen I, Stiernholm P, Jensen K, Olesen J, Sundler F (1986) Peptide-containing nerve fibres in human extracranial tissue: a morphological basis for neuropeptide involvement in extracranial pain? Pain 27:391–399

Voss T, Lipton RB, Dodick DW, Dupre N, Ge JY, Bachman R, Assaid C, Aurora SK, Michelson D (2016) A phase IIb randomized, double-blind, placebo-controlled trial of ubrogepant for the acute treatment of migraine. Cephalalgia 36:887–898. https://doi.org/10.1177/0333102416653233

Walker CS, Conner AC, Poyner DR, Hay DL (2010) Regulation of signal transduction by calcitonin gene-related peptide receptors. Trends Pharmacol Sci 31:476–483. https://doi.org/10.1016/j.tips.2010.06.006

Walker CS, Eftekhari S, Bower RL, Wilderman A, Insel PA, Edvinsson L, Waldvogel HJ, Jamaluddin MA, Russo AF, Hay DL (2015) A second trigeminal CGRP receptor: function and expression of the AMY1 receptor. Ann Clin Transl Neurol 2:595–608. https://doi.org/10.1002/acn3.197

Williamson DJ, Hargreaves RJ, Hill RG, Shepheard SL (1997) Sumatriptan inhibits neurogenic vasodilation of dural blood vessels in the anaesthetized rat – intravital microscope studies. Cephalalgia 17:525–531. https://doi.org/10.1046/j.1468-2982.1997.1704525.x

Wysham C, Lush C, Zhang B, Maier H, Wilhelm K (2008) Effect of pramlintide as an adjunct to basal insulin on markers of cardiovascular risk in patients with type 2 diabetes. Curr Med Res Opin 24:79–85. https://doi.org/10.1185/030079908X253537

CGRP Receptor Biology: Is There More Than One Receptor?

Debbie L. Hay

Contents

Abstract

Calcitonin gene-related peptide (CGRP) has many reported pharmacological actions. Can a single receptor explain all of these? This chapter outlines the molecular nature of reported CGRP binding proteins and their pharmacology. Consideration of whether CGRP has only one or has more receptors is important because of the key role that this peptide plays in migraine. It is widely thought that the calcitonin receptor-like receptor together with receptor activity-modifying protein 1 (RAMP1) is the only relevant receptor for CGRP. However, some closely related receptors also have high affinity for CGRP and it is still plausible that these play a role in CGRP biology, and in migraine. The calcitonin receptor/RAMP1 complex, which is currently called the AMY_1 receptor, seems to be the most likely candidate but more investigation is needed to determine its role.

Keywords

AMY_1 · CGRP · CGRP receptor · Migraine · RAMP1

D. L. Hay (✉)
School of Biological Sciences, The University of Auckland, Auckland, New Zealand
e-mail: dl.hay@auckland.ac.nz

© Springer International Publishing AG, part of Springer Nature 2018   13
S. D. Brain, P. Geppetti (eds.), *Calcitonin Gene-Related Peptide (CGRP) Mechanisms*,
Handbook of Experimental Pharmacology 255, https://doi.org/10.1007/164_2018_131

1   Introduction

Calcitonin (CT), amylin, CT gene-related peptide (CGRP), adrenomedullin (AM) and adrenomedullin 2/intermedin (AM2) comprise the major members of the CT family of peptides. Although the members of this family share only low amino acid sequence homology, they have common structural features. Each peptide has two cysteine residues close to its N-terminus which form a disulphide bridge; this region is critical for biological activity in all cases. An α-helical region is present towards the N-terminus of each of the peptides and an amidated residue at the carboxy terminus is also a family trait.

CGRP has long been known to trigger receptor-dependent cellular signalling. However, the pharmacology and molecular identity of the receptor(s) is not "clean". This chapter will describe the molecular composition of known CGRP binding proteins, the activity of CGRP in relation to related peptides within the same family and in broad terms how this information reconciles with CGRP actions in cells and tissues. An important consideration is the behaviour of receptors from different model species. In order to understand CGRP receptors, it is also important to consider related peptides and their receptors because of the substantial functional overlap between their actions.

2   Proteins with Affinity for Calcitonin Gene-Related Peptide

The first molecular candidates for AM and CGRP receptors were proposed in 1995. One of these was a rat receptor that was reportedly responsive to AM. This G protein-coupled receptor (GPCR), receptor named L1 (or G10d), had a similar pattern of mRNA expression to that of ^{125}I-AM binding in rat tissues and when transfected into Cos-7 cells, elevated cAMP with an approximate EC_{50} of 7 nM in response to AM (Kapas et al. 1995). Also in 1995, the same group reported that a then orphan canine GPCR, RDC-1 (which shares approximately 30% sequence identity with L1), was a CGRP receptor as transfection into Cos-7 cells yielded cAMP production in response to CGRP, which could be blocked by $CGRP_{8-37}$. AM could also stimulate cAMP levels in cells transfected with RDC-1 with potency that might be expected for a CGRP receptor (Kapas and Clark 1995). In 1997, a potential human AM receptor with 73% amino acid homology to L1 was reported but the receptor had a different expression pattern to the rat receptor (Hanze et al. 1997). Attempts to further characterise these receptors as AM and CGRP receptors, respectively, have been unsuccessful (Kennedy et al. 1998; McLatchie et al. 1998). No binding of rat or human ^{125}I-AM could be found following transfection of either the putative human or rat AM receptor sequences. Furthermore, expression of RDC-1 in *Xenopus* oocytes and human embryonic kidney (HEK) 293 cells did not change the cellular response to CGRP (McLatchie et al. 1998). A report describing correlations between AM and CGRP receptor binding and their proposed molecular counterparts found no correlation between CGRP binding and RDC-1 mRNA in rat tissues (Chakravarty et al. 2000). According to the guidelines produced by the relevant nomenclature committee of the International Union of Pharmacology, these receptors are not considered to be candidate AM or CGRP receptors (Hay et al. 2018; Poyner et al. 2002). The human receptor corresponding to RDC-1 is now known as atypical chemokine receptor 3 and thought to be a decoy receptor; it was formerly called GPR159 or CXCR7 (see www.guidetopharmacology.org). However, another link between AM and this receptor has been found, whereby this receptor has been proposed as a decoy receptor for this peptide (Klein et al. 2014). More work is needed to determine how significant this finding is to AM biology. L1 is known as GPR182 and is considered to be a class A orphan GPCR.

In addition to L1 and RDC-1, a further GPCR, belonging to the class B sub-group and sharing no significant amino acid sequence identity to those receptors, was cloned from rat pulmonary blood vessels (Njuki et al. 1993). This receptor has ~50% overall amino acid sequence identity with the CT receptor (CTR) and was thus named the CT receptor-like receptor (Njuki et al. 1993). The abbreviation CLR is now used for simplicity (Hay et al. 2018). The human isoform of CLR was identified in 1995 from cerebellum and is closely related to rat CLR (Fluhmann et al. 1995). CLR or CLR-like receptors have been cloned from many other species, including porcine, bovine, mouse and flounder (Aiyar et al. 2002; Elshourbagy et al. 1998; Miyauchi et al. 2002; Sekiguchi et al. 2016; Suzuki et al. 2000). Early attempts to find a ligand for CLR were unsuccessful (Fluhmann et al. 1995). However, when human CLR was transfected into HEK293 cells a clear CGRP receptor pharmacology was observed (Aiyar et al. 1996). These studies differed from previous ones in the cell type used for transfection. Initially, Cos-7 cells had been used. The results of Aiyar and colleagues were confirmed with rat CLR and later porcine CLR was also identified as a CGRP receptor in HEK293 cells (Elshourbagy et al. 1998; Han et al. 1997). These studies led to the pertinent question: what was the factor in HEK293 cells that allowed CLR to act as a fully functional receptor for CGRP (Han et al. 1997)?

In 1998, Foord and colleagues provided new insight into the way that GPCRs and their pharmacology can be regulated (McLatchie et al. 1998). A new family of single transmembrane domain proteins termed receptor activity-modifying proteins (RAMPs) were discovered. It was found that these proteins were required for functional expression of CLR at the cell surface, explaining why it had been so difficult to find binding or function when this receptor was transfected into cells not expressing RAMPs (Fluhmann et al. 1995; Han et al. 1997; McLatchie et al. 1998).

Three human RAMPs have been cloned; RAMP1, RAMP2 (both cloned from SK-N-MC cells) and RAMP3 (cloned from human spleen) (McLatchie et al. 1998). Their amino acid sequences are 31% identical and 56% similar to one another and the respective rat and mouse (and other species) RAMPs have been cloned in addition to human (Husmann et al. 2000; McLatchie et al. 1998; Nagae et al. 2000; Oliver et al. 2001; Miyauchi et al. 2002). The structures of the extracellular domain region of CLR in complex with RAMP1 and RAMP2 are now known, providing useful insights into their mechanism of action (Booe et al. 2015; Kusano et al. 2012; ter Haar et al. 2010).

The discovery that RAMPs are essential for the functionality of CLR explains why this receptor was an orphan until this time. Of considerable interest was the discovery that the receptors generated by CLR/RAMP complexes recapitulated CGRP and AM pharmacology that had previously been described in various tissues and cell lines (Foord et al. 1999; Fraser et al. 1999; McLatchie et al. 1998). The discovery of RAMPs finally assigned firm molecular entities as CGRP and AM receptors, laying the foundation of all subsequent work. Co-expression of human CLR with RAMP1 constituted a CGRP receptor and co-expression of CLR with RAMP2 or RAMP3 yielded AM receptors (McLatchie et al. 1998). The AM receptors formed by CLR with RAMP2 or RAMP3 are defined as AM_1 and AM_2 subtypes, respectively (Hay et al. 2018); see Table 1. However, the precise nature of the pharmacology observed can be dependent on the species of the RAMP/CLR proteins. For example, rodent RAMP3-based receptors have a more mixed AM/CGRP receptor phenotype than the human equivalents (Hay et al. 2003; Husmann et al. 2000; McLatchie et al. 1998). This is covered further below. Table 1 summarises the receptors that are formed from CLR/RAMP complex formation.

In 1996, another protein that conferred CGRP responsiveness in *Xenopus* oocytes was described. This 146-amino acid protein was cloned from guinea pig organ of Corti and does not share the classical seven transmembrane domains of the other receptors described for the CT family of peptides (Luebke et al. 1996). This protein, called "receptor component protein" (RCP) does not form a direct binding site for CGRP but appears to be a key component of the CGRP receptor signalling system (Luebke et al. 1996). RCP acts as a peripheral membrane protein that facilitates receptor signalling. The physiological significance of this protein is still unclear, and compared to the CLR/RAMP1 complex itself, it is poorly studied.

Table 1 Summary of the molecular components and agonist pharmacology for the CT family of peptides (Bailey et al. 2012; Halim and Hay 2012; Hay et al. 2002, 2003, 2018; Husmann et al. 2000; Miyauchi et al. 2002; Oliver et al. 2001)

Receptor name	Molecular components	Human agonist pharmacology	Rodent agonist pharmacology[a]
CGRP	CLR + RAMP1	CGRP>AM = AM2	CGRP>AM = AM2
AM_1	CLR + RAMP2	AM>AM2 ≥ CGRP	AM> > CGRP
AM_2	CLR + RAMP3	AM = AM2 > CGRP	AM≥AM2 ≥ CGRP
CT	Calcitonin receptor	CT > Amy ≥ CGRP	CT > Amy = CGRP
AMY_1	Calcitonin receptor + RAMP1	Amy = CGRP	Amy = CGRP
AMY_2	Calcitonin receptor + RAMP2	Poorly defined	Poorly defined
AMY_3	Calcitonin receptor + RAMP3	Amy > CGRP	Amy = CGRP

[a]Some of these studies use a mixture of rat and mouse receptor components, making the pharmacology of discrete rodent receptors difficult to define

Soon after the discovery of RAMPs, it was identified that these proteins could interact with CTR, in addition to CLR. RAMPs also change CTR pharmacology (Hay et al. 2018). Here, there is an alteration in CGRP and amylin pharmacology, depending on whether RAMP1, RAMP2 or RAMP3 is expressed. Human CTR with human RAMP1 (the AMY_1 receptor) forms a dual high affinity receptor for both amylin and CGRP, whereas the human RAMP2 and RAMP3 complexes (AMY_2 and AMY_3, respectively) are high affinity amylin receptors. In a similar manner to CLR complexes, however, species plays a role in pharmacology.

3 Pharmacology of Calcitonin Gene-Related Peptide-Responsive Receptors

CLR/RAMP1 is now considered to be the canonical CGRP receptor and has high affinity for CGRP and around 10- to 20-fold lower affinity for AM and AM2 (Hay et al. 2018). Amylin and CT are weaker agonists at this complex. Amylin can activate this receptor but high nanomolar concentrations of the peptide are required to do this at the human receptor. At the rat receptor, amylin is more potent (Walker et al. 2015). Several antagonists have been developed that act at this receptor. These include fragments of CGRP itself, such as $CGRP_{8-37}$, or shorter C-terminal fragments (Watkins et al. 2013). A number of small molecules with high affinity for this receptor have also been identified as part of migraine drug development programmes. Examples of these are olcegepant, telcagepant, atogepant, ubrogepant and rimagepant. The pharmacology of several but not all of these is in the public domain. In general, these have high affinity for CLR/RAMP1 and lower but appreciable affinity for CTR/RAMP1, and very low or not measurable affinity for CTR alone or RAMP2- or RAMP3-based complexes with CLR or CTR (Moore and Salvatore 2012). However, it is important to note that cross-receptor pharmacology is only readily available for olcegepant and telcagepant, and it is not known for certain how rimagepant, atogepant and ubrogepant behave at all the different receptor complexes. Thus, it is only assumed that their pharmacology will be similar to structurally related compounds where the data are fully reported.

CTR/RAMP1 has also long been recognised to have high affinity for CGRP. This was first reported in the earliest studies of this receptor complex but has since been replicated many times (Christopoulos et al. 1999; Hay et al. 2005, 2018; Hay and Walker 2017; Kuwasako et al. 2004; Leuthauser et al. 2000; Tilakaratne et al. 2000). Although current nomenclature calls this complex the AMY_1 receptor due to its high affinity for amylin, CGRP equals amylin in potency/affinity and on this basis alone could be considered as a dual amylin/CGRP receptor (Hay et al. 2018). This naming convention has probably caused this receptor to be less studied as a CGRP receptor. It is important that this complex is studied both in terms of CGRP and amylin biology to determine its cognate ligand(s) in vivo because it is not known how important this receptor is for amylin biology either (Hay et al. 2015). Recent data has shown that CTR/RAMP1 may be activated by CGRP endogenously because it has been found in the trigeminovascular system where CGRP is abundant (Walker et al. 2015). There are some antagonists of CTR/RAMP complexes (e.g. AC187, AC413 and salmon CT_{8-32}) but none that are selective for CTR/RAMP1; hence, these are only of limited use in characterising this receptor (Hay et al. 2005, 2015). As with this whole family because there is significant cross-reactivity of most agonist and antagonist ligands across two or more receptors, it is recommended that multiple tools are used to probe the identity and function of the receptors (Hay et al. 2018).

It is generally thought that CGRP has little activity at other CLR and CTR complexes. However, this is not necessarily true. Rodent CLR/RAMP3 (AM_2) and CTR/RAMP3 (AMY_3) receptors have higher affinity for CGRP than their human counterparts (Bailey et al. 2012; Halim and Hay 2012; Hay et al. 2003; Husmann et al. 2000). CGRP potency/affinity can almost equal that of AM at the rat AM_2 receptor (Hay et al. 2003). It is not known whether this is something that only occurs in isolated cells because more work is needed in this area. Hence, in terms of in vitro receptor pharmacology CGRP has high affinity at CLR/RAMP1 and CTR/RAMP1 across species and can have high affinity for RAMP3-based CLR and CTR receptors in a species-dependent manner. Thus, at the AMY_3 and AM_2 receptors, CGRP has higher affinity at rodent than human receptors (Bailey et al. 2012; Halim and Hay 2012; Hay et al. 2003). CGRP also has affinity at RAMP2-based receptors (Hay et al. 2018; Husmann et al. 2000). However, it is important to point out that there are not many studies of non-human receptors and many of these use mixed-species components. Further work is needed to more deeply characterise CGRP responses across these receptor complexes in different species, including other model species that are important for drug development such as dog and other primates. When considering what receptor CGRP may be acting through, there are actually quite a few possibilities and it should not be forgotten that complex CGRP pharmacology has been known for several decades (Dennis et al. 1990; Juaneda et al. 2000; Poyner 1995). The CLR/RAMP1 complex is clearly an important CGRP receptor but there is much evidence suggesting that this may not be the only relevant molecular entity responsible for the actions of CGRP.

A further factor to consider for the pharmacology of CGRP receptors is the potential differences that may be found when considering different signalling pathways. Most work is done measuring cAMP and to a more limited extent radioligand binding, which has given the pharmacological nomenclature that we have to date. However, distinct profiles are certainly possible when considering the plethora of other signalling pathways that CGRP can activate and there is emerging evidence that this may need to be considered to a greater extent both for agonists and antagonists (Walker et al. 2010, 2017; Weston et al. 2016). Moreover, for CTR there are also splice variants to consider which differ in expression, signalling and pharmacological profiles with or without RAMPs (Furness et al. 2012; Qi et al. 2013).

4   Calcitonin Gene-Related Peptide Receptors in Migraine

As will be evident from other chapters in this book, the CLR/RAMP1 complex is widely considered to be the relevant complex for migraine. This is the receptor that small molecule drugs have the highest affinity for and the complex to which an antibody drug has been targeted. This is also the receptor that most work has been done on to identify its localisation in migraine-relevant tissues. Does this mean that this is the only relevant receptor to CGRP in general or more specifically to CGRP in migraine? At this time, the answer has to be "no" because there is insufficient evidence to rule out a role for other complexes that bind CGRP with high affinity. As examples, this is because many small molecule antagonists that have been designed to target CLR/RAMP1 can also block CGRP activity at CTR/RAMP1 (Hay et al. 2006; Moore and Salvatore 2012; Salvatore et al. 2010; Walker et al. 2017), and because CTR/RAMP1 expression has hardly been studied in tissues so it is not known whether this receptor is present or absent at sites relevant to CGRP action. The monoclonal antibodies in clinical development bind either to the CGRP peptide itself or to the CLR/RAMP1 complex. In time, the clinical data may show some separation either in efficacy or differences in adverse events between these mechanisms. At this stage, it is too early to say whether the targeting of only CLR/RAMP1 by the Amgen/Novartis antibody, assuming it has no ability to bind to CTR/RAMP1 [there is limited data in the public domain (Shi et al. 2016)], shows any benefit over targeting the ligand, which presumably impedes activity at all CGRP-responsive receptors. A molecule that selectively targets CTR/RAMP1 is needed to firmly establish whether or not this receptor plays a role in migraine.

In conclusion, because the human CTR/RAMP1 AMY_1 receptor has high affinity for CGRP and has been reported in migraine-relevant tissue, this complex should certainly be considered as a possible player in CGRP action in migraine in addition to the canonical CLR/RAMP1 CGRP receptor.

References

Aiyar N, Rand K, Elshourbagy NA, Zeng Z, Adamou JE, Bergsma DJ, Li Y (1996) A cDNA encoding the calcitonin gene-related peptide type 1 receptor. J Biol Chem 271(19):11325–11329

Aiyar N, Disa J, Ao Z, Xu D, Surya A, Pillarisetti K, Parameswaran N, Gupta SK, Douglas SA, Nambi P (2002) Molecular cloning and pharmacological characterization of bovine calcitonin receptor-like receptor from bovine aortic endothelial cells. Biochem Pharmacol 63 (11):1949–1959

Bailey RJ, Walker CS, Ferner AH, Loomes KM, Prijic G, Halim A, Whiting L, Phillips ARJ, Hay DL (2012) Pharmacological characterisation of rat amylin receptors: implications for the identification of amylin receptor subtypes. Br J Pharmacol 166(1):151–167. https://doi.org/10.1111/j.1476-5381.2011.01717.x

Booe JM, Walker CS, Barwell J, Kuteyi G, Simms J, Jamaluddin MA, Warner ML, Bill RM, Harris PW, Brimble MA, Poyner DR, Hay DL, Pioszak AA (2015) Structural basis for receptor activity-modifying protein-dependent selective peptide recognition by a G protein-coupled receptor. Mol Cell 58(6):1040–1052. https://doi.org/10.1016/j.molcel.2015.04.018

Chakravarty P, Suthar TP, Coppock HA, Nicholl CG, Bloom SR, Legon S, Smith DM (2000) CGRP and adrenomedullin binding correlates with transcript levels for calcitonin receptor-like receptor (CRLR) and receptor activity modifying proteins (RAMPs) in rat tissues. Br J Pharmacol 130(1):189–195

Christopoulos G, Perry K, Morfis M, Tilakaratne N, Gao Y, Fraser NJ, Main MJ, Foord SM, Sexton PM (1999) Multiple amylin receptors arise from receptor-activity-modifying-proteins interaction with the calcitonin receptor gene product. Mol Pharmacol 56:235–242

Dennis T, Fournier A, Cadieux A, Pomerleau F, Jolicoeur FB, St. Pierre S, Quirion R (1990) hCGRP8-37, a calcitonin gene-related peptide antagonist revealing calcitonin gene-related peptide receptor heterogeneity in brain and periphery. J Pharmacol Exp Ther 254(1):123–128

Elshourbagy NA, Adamou JE, Swift AM, Disa J, Mao J, Ganguly S, Bergsma DJ, Aiyar N (1998) Molecular cloning and characterization of the porcine calcitonin gene-related peptide receptor. Endocrinology 139(4):1678–1683

Fluhmann B, Muff R, Hunziker W, Fischer JA, Born W (1995) A human orphan calcitonin receptor-like structure. Biochem Biophys Res Commun 206(1):341–347

Foord SM, Wise A, Brown J, Main MJ, Fraser NJ (1999) The N-terminus of RAMPs is a critical determinant of the glycosylation state and ligand binding of calcitonin receptor-like receptor. Biochem Soc Trans 27(4):535–539

Fraser NJ, Wise A, Brown J, McLatchie LM, Main MJ, Foord SM (1999) The amino terminus of receptor activity modifying proteins is a crucial determinant of glycosylation state and ligand binding of calcitonin-receptor-like-receptor. Mol Pharmacol 55:1054–1059

Furness SGB, Wootten D, Christopoulos A, Sexton PM (2012) Consequences of splice variation on secretin family G protein-coupled receptor function. Br J Pharmacol 166(1):98–109. https://doi.org/10.1111/j.1476-5381.2011.01571.x

Halim A, Hay DL (2012) The role of glutamic acid 73 in adrenomedullin interactions with rodent AM2 receptors. Peptides 36(1):137–141. https://doi.org/10.1016/j.peptides.2012.04.011

Han ZQ, Coppock HA, Smith DM, Van Noorden S, Makgoba MW, Nicholl CG, Legon S (1997) The interaction of CGRP and adrenomedullin with a receptor expressed in the rat pulmonary vascular endothelium. J Mol Endocrinol 18(3):267–272

Hanze J, Dittrich K, Dotsch J, Rascher W (1997) Molecular cloning of a novel human receptor gene with homology to the rat adrenomedullin receptor and high expression in heart and immune system. Biochem Biophys Res Commun 240(1):183–188

Hay DL, Walker CS (2017) CGRP and its receptors. Headache 57(4):625–636. https://doi.org/10.1111/head.13064

Hay DL, Howitt SG, Conner AC, Doods H, Schindler M, Poyner DR (2002) A comparison of the actions of BIBN4096BS and CGRP8-37 on CGRP and adrenomedullin receptors expressed on SK-N-MC, L6, Col 29 and Rat 2 cells. Br J Pharmacol 137(1):80

Hay DL, Howitt SG, Conner AC, Schindler M, Smith DM, Poyner DR (2003) CL/RAMP2 and CL/RAMP3 produce pharmacologically distinct adrenomedullin receptors: a comparison of effects of adrenomedullin22-52, CGRP8-37 and BIBN4096BS. Br J Pharmacol 140(3):477–486

Hay DL, Christopoulos G, Christopoulos A, Poyner DR, Sexton PM (2005) Pharmacological discrimination of calcitonin receptor: receptor activity-modifying protein complexes. Mol Pharmacol 67(5):1655–1665. https://doi.org/10.1124/mol.104.008615

Hay DL, Christopoulos G, Christopoulos A, Sexton PM (2006) Determinants of 1-piperidinecarboxamide, N-[2-[[5-amino-l-[[4-(4-pyridinyl)-l-piperazinyl]carbonyl]pentyl]amino]-1-[(3,5-d ibromo-4-hydroxyphenyl)methyl]-2-oxoethyl]-4-(1,4-dihydro-2-oxo-3(2H)-quinazoliny l) (BIBN4096BS) affinity for calcitonin gene-related peptide and amylin receptors – the role of receptor activity modifying protein 1. Mol Pharmacol 70(6):1984–1991. https://doi.org/10.1124/mol.106.027953

Hay DL, Chen S, Lutz TA, Parkes DG, Roth JD (2015) Amylin: pharmacology, physiology, and clinical potential. Pharmacol Rev 67(3):564–600. https://doi.org/10.1124/pr.115.010629

Hay DL, Garelja ML, Poyner DR, Walker CS (2018) Update on the pharmacology of calcitonin/ CGRP family of peptides: IUPHAR review 25. Br J Pharmacol 175(1):3–17. https://doi.org/10. 1111/bph.14075

Husmann K, Sexton PM, Fischer JA, Born W (2000) Mouse receptor-activity-modifying proteins 1, -2 and -3: amino acid sequence, expression and function. Mol Cell Endocrinol 162 (1–2):35–43

Juaneda C, Dumont Y, Quirion R (2000) The molecular pharmacology of CGRP and related peptide receptor subtypes. Trends Pharmacol Sci 21:432–438

Kapas S, Clark AJ (1995) Identification of an orphan receptor gene as a type 1 calcitonin gene-related peptide receptor. Biochem Biophys Res Commun 217(3):832–838

Kapas S, Catt KJ, Clark AJ (1995) Cloning and expression of cDNA encoding a rat adrenomedullin receptor. J Biol Chem 270(43):25344–25347

Kennedy SP, Sun D, Oleynek JJ, Hoth CF, Kong J, Hill RJ (1998) Expression of the rat adrenomedullin receptor or a putative human adrenomedullin receptor does not correlate with adrenomedullin binding or functional response. Biochem Biophys Res Commun 244:832–837

Klein KR, Karpinich NO, Espenschied ST, Willcockson HH, Dunworth WP, Hoopes SL, Kushner EJ, Bautch VL, Caron KM (2014) Decoy receptor CXCR7 modulates adrenomedullin-mediated cardiac and lymphatic vascular development. Dev Cell 30(5):528–540. https://doi.org/10.1016/ j.devcel.2014.07.012

Kusano S, Kukimoto-Niino M, Hino N, Ohsawa N, Okuda K-I, Sakamoto K, Shirouzu M, Shindo T, Yokoyama S (2012) Structural basis for extracellular interactions between calcitonin receptor-like receptor and receptor activity-modifying protein 2 for adrenomedullin-specific binding. Protein Sci 21(2):199–210. https://doi.org/10.1002/pro.2003

Kuwasako K, Cao Y-N, Nagoshi Y, Tsuruda T, Kitamura K, Eto T (2004) Characterization of the human calcitonin gene-related peptide receptor subtypes associated with receptor activity-modifying proteins. Mol Pharmacol 65(1):207–213

Leuthauser K, Gujer R, Aldecoa A, McKinney RA, Muff R, Fischer JA, Born W (2000) Receptor-activity-modifying protein 1 forms heterodimers with two G-protein-coupled receptors to define ligand recognition. Biochem J 351:347–351

Luebke AE, Dahl GP, Roos BA, Dickerson IM (1996) Identification of a protein that confers calcitonin gene-related peptide responsiveness to oocytes by using a cystic fibrosis transmembrane conductance regulator assay. Proc Natl Acad Sci U S A 93(8):3455–3460

McLatchie LM, Fraser NJ, Main MJ, Wise A, Brown J, Thompson N, Solari R, Lee MG, Foord SM (1998) RAMPs regulate the transport and ligand specificity of the calcitonin-receptor-like receptor. Nature 393(6683):333–339

Miyauchi K, Tadotsu N, Hayashi T, Ono Y, Tokoyoda K, Tsujikawa K, Yamamoto H (2002) Molecular cloning and characterization of mouse calcitonin gene-related peptide receptor. Neuropeptides 36(1):22–33

Moore EL, Salvatore CA (2012) Targeting a family B GPCR/RAMP receptor complex: CGRP receptor antagonists and migraine. Br J Pharmacol 166(1):66–78. https://doi.org/10.1111/j. 1476-5381.2011.01633.x

Nagae T, Mukoyama M, Sugawara A, Mori K, Yahata K, Kasahara M, Suganami T, Makino H, Fujinaga Y, Yoshioka T, Tanaka K, Nakao K (2000) Rat receptor-activity-modifying proteins (RAMPs) for adrenomedullin/CGRP receptor: cloning and upregulation in obstructive nephropathy. Biochem Biophys Res Commun 270:89–93

Njuki F, Nicholl CG, Howard A, Mak JCW, Barnes PJ, Girgis SI, Legon S (1993) A new calcitonin-receptor-like sequence in rat pulmonary blood vessels. Clin Sci 85:385–388

Oliver KR, Kane SA, Salvatore CA, Mallee JJ, Kinsey AM, Koblan KS, Keyvan-Fouladi N, Heavens RP, Wainwright A, Jacobson M, Dickerson IM, Hill RG (2001) Cloning, characterization and central nervous system distribution of receptor activity modifying proteins in the rat. Eur J Neurosci 14(4):618–628. https://doi.org/10.1046/j.0953-816x.2001.01688.x

Poyner D (1995) Pharmacology of receptors for calcitonin gene-related peptide and amylin. Trends Pharmacol Sci 16(12):424–428

Poyner DR, Sexton PM, Marshall I, Smith DM, Quirion R, Born W, Muff R, Fischer JA, Foord SM (2002) International Union of Pharmacology. XXXII. The mammalian calcitonin gene-related peptides, Adrenomedullin, amylin, and calcitonin receptors. Pharmacol Rev 54(2):233–246

Qi T, Dong M, Watkins HA, Wootten D, Miller LJ, Hay DL (2013) Receptor activity-modifying protein-dependent impairment of calcitonin receptor splice variant Δ(1–47)hCT(a) function. Br J Pharmacol 168(3):644–657. https://doi.org/10.1111/j.1476-5381.2012.02197.x

ter Haar E, Koth CM, Abdul-Manan N, Swenson L, Coll JT, Lippke JA, Lepre CA, Garcia-Guzman M, Moore JM (2010) Crystal structure of the ectodomain complex of the CGRP receptor, a class-B GPCR, reveals the site of drug antagonism. Structure 18(9):1083–1093. https://doi.org/10.1016/j.str.2010.05.014

Salvatore CA, Moore EL, Calamari A, Cook JJ, Michener MS, O'Malley S, Miller PJ, Sur C, Williams DL Jr, Zeng Z, Danziger A, Lynch JJ, Regan CP, Fay JF, Tang YS, Li CC, Pudvah NT, White RB, Bell IM, Gallicchio SN, Graham SL, Selnick HG, Vacca JP, Kane SA (2010) Pharmacological properties of MK-3207, a potent and orally active calcitonin gene-related peptide receptor antagonist. J Pharmacol Exp Ther 333(1):152–160. https://doi.org/10.1124/jpet.109.163816

Sekiguchi T, Kuwasako K, Ogasawara M, Takahashi H, Matsubara S, Osugi T, Muramatsu I, Sasayama Y, Suzuki N, Satake H (2016) Evidence for conservation of the calcitonin superfamily and activity-regulating mechanisms in the basal chordate Branchiostoma floridae: insights into the molecular and functional evolution in chordates. J Biol Chem 291(5):2345–2356. https://doi.org/10.1074/jbc.M115.664003

Shi L, Lehto SG, Zhu DXD, Sun H, Zhang J, Smith BP, Immke DC, Wild KD, Xu C (2016) Pharmacologic characterization of AMG 334, a potent and selective human monoclonal antibody against the calcitonin gene-related peptide receptor. J Pharmacol Exp Ther 356 (1):223–231. https://doi.org/10.1124/jpet.115.227793

Suzuki N, Suzuki T, Kurokawa T (2000) Cloning of a calcitonin gene-related peptide receptor and a novel calcitonin receptor-like receptor from the gill of the flounder, Paralichthys olivaceus. Gene 244:81–88

Tilakaratne N, Christopoulos G, Zumpe ET, Foord SM, Sexton PM (2000) Amylin receptor phenotypes derived from human calcitonin receptor/RAMP coexpression exhibit pharmacological differences dependent on receptor isoform and host cell environment. J Pharmacol Exp Ther 294:61–72

Walker CS, Conner AC, Poyner DR, Hay DL (2010) Regulation of signal transduction by calcitonin gene-related peptide receptors. Trends Pharmacol Sci 31(10):476–483. https://doi.org/10.1016/j.tips.2010.06.006

Walker CS, Eftekhari S, Bower RL, Wilderman A, Insel PA, Edvinsson L, Waldvogel HJ, Jamaluddin MA, Russo AF, Hay DL (2015) A second trigeminal CGRP receptor: function and expression of the AMY1 receptor. Ann Clin Transl Neurol 2(6):595–608. https://doi.org/10.1002/acn3.197

Walker CS, Raddant AC, Woolley MJ, Russo AF, Hay DL (2017) CGRP receptor antagonist activity of olcegepant depends on the signalling pathway measured. Cephalalgia 38 (3):437–451. https://doi.org/10.1177/0333102417691762

Watkins HA, Rathbone DL, Barwell J, Hay DL, Poyner DR (2013) Structure-activity relationships for alpha-calcitonin gene-related peptide. Br J Pharmacol 170(7):1308–1322. https://doi.org/10.1111/bph.12072

Weston C, Winfield I, Harris M, Hodgson R, Shah A, Dowell SJ, Mobarec JC, Woodlock DA, Reynolds CA, Poyner DR, Watkins HA, Ladds G (2016) Receptor activity-modifying protein-directed G protein signaling specificity for the calcitonin gene-related peptide family of receptors. J Biol Chem 291(42):21925–21944. https://doi.org/10.1074/jbc.M116.751362

The Structure of the CGRP and Related Receptors

John Simms, Sarah Routledge, Romez Uddin, and David Poyner

Contents

Abstract

The canonical CGRP receptor is a complex between calcitonin receptor-like receptor (CLR), a family B G-protein-coupled receptor (GPCR) and receptor activity-modifying protein 1 (RAMP1). A third protein, receptor component protein (RCP) is needed for coupling to Gs. CGRP can interact with other RAMP–receptor complexes, particularly the AMY1 receptor formed between the calcitonin receptor (CTR) and RAMP1. Crystal structures are available for the binding of $CGRP_{27-37}$ [D^{31},P^{34},F^{35}] to the extracellular domain (ECD) of CLR and RAMP1; these show that extreme C-terminal amide of CGRP interacts with W84 of RAMP1 but the rest of the analogue interacts with CLR. Comparison with the crystal structure of a fragment of the allied peptide adrenomedullin

J. Simms
School of Life and Health Science, Aston University, Birmingham, UK

Coventry University, Coventry, UK

S. Routledge · R. Uddin · D. Poyner (✉)
School of Life and Health Science, Aston University, Birmingham, UK
e-mail: D.R.Poyner@aston.ac.uk

© Springer International Publishing AG, part of Springer Nature 2018   23
S. D. Brain, P. Geppetti (eds.), *Calcitonin Gene-Related Peptide (CGRP) Mechanisms*,
Handbook of Experimental Pharmacology 255, https://doi.org/10.1007/164_2018_132

bound to the ECD of CLR/RAMP2 confirms the importance of the interaction of the ligand C-terminus and the RAMP in determining pharmacology specificity, although the RAMPs probably also have allosteric actions. A cryo-electron microscope structure of calcitonin bound to the full-length CTR associated with Gs gives important clues as to the structure of the complete receptor and suggests that the N-terminus of CGRP makes contact with $His^{5.40b}$, high on TM5 of CLR. However, it is currently not known how the RAMPs interact with the TM bundle of any GPCR. Major challenges remain in understanding how the ECD and TM domains work together to determine ligand specificity, and how G-proteins influence this and the role of RCP. It seems likely that allosteric mechanisms are particularly important as are the dynamics of the receptors.

Keywords

Allostery · Amylin · Calcitonin · Cryo-electron microscopy · Crystallography · Family B G-protein-coupled receptor · Molecular dynamics

1   Introduction to the CGRP Receptors

The CGRP receptor as defined by IUPHAR is the complex between calcitonin receptor-like receptor (CLR) and receptor activity-modifying protein 1 (RAMP1) (Hay et al. 2017). It was the first receptor to be identified as a complex between a G-protein-coupled receptor (GPCR) and a RAMP (McLatchie et al. 1998). However, the peptide also has a high affinity at the AMY1 receptor formed between the calcitonin receptor (CTR) and RAMP1; its potency at this receptor is equal to that of amylin and it may be an endogenous agonist here (Walker et al. 2015). It has modest affinity at the AM2 and AMY3 receptors (CLR/RAMP2 and CTR/RAMP3, respectively) and can stimulate the AM1 and AMY2 receptors at pharmacological concentrations. In the older literature, there was much talk of CGRP1 and CGRP2 receptors; the CGRP1 receptor corresponds to the CLR/RAMP1 complex whereas it is likely that much of the pharmacology ascribed to the CGRP2 receptor in fact corresponds to the AMY1 receptor, perhaps with some contribution from the AM2 and AMY3 receptors (Hay et al. 2008). Most analysis has been done on the CGRP receptor and this will be the focus of the current chapter.

Both CLR and CTR are class B GPCRs. The three RAMPs each have an N-terminus of around 100–120 amino acids, a single transmembrane domain (TMD) and a C-terminus of around ten residues; their structures have been recently reviewed elsewhere (Hay and Pioszak 2016), as has the structure–activity relationship for CGRP (Watkins et al. 2013). In brief, for CLR the RAMPs have two main functions; they translocate CLR to the cell surface and also create the binding pocket for the endogenous peptides and also some non-peptide antagonists (McLatchie et al. 1998; Hay and Pioszak 2016).

Activation of CLR follows the two-domain model found in other class B GPCRs, where this is achieved by binding of the peptide N-terminus to the TMD of the receptor. The C-terminus of the peptide binds to the extracellular domain (ECD) of the receptor, contributing to the overall affinity of the peptide. CLR couples

Table 1 Summary of crystal structures of the extracellular domains of CTR and CLR/RAMP complexes with bound ligands

RAMP	GPCR	Ligand	RSCB ID/Reference
$RAMP1_{26-117}$	CLR_{22-133}	Telcagepant	3N7R, ter Haar et al. (2010)
$RAMP1_{26-117}$	CLR_{22-133}	Olcegepant	3N7S, ter Haar et al. (2010)
$MBP\text{-}RAMP1_{24-111}\text{-}(GSA)_3\text{-}$ $CLR_{29-144}\text{-}(H)_6$		$CGRP_{27-37}\,[D^{31},P^{34},F^{35}]$	4RWG, Booe et al. (2015)
$MBP\text{-}RAMP2[L106R]_{55-140}\text{-}$ $(GSA)_3\text{-}CLR_{29-144}\text{-}(H)_6$		AM_{25-52}	4RWF, Booe et al. (2015)
–	$H_6\text{-}CTR_{25-144}$	$[BrPhe^{22}]sCT_{8-32}$	5II0, Johansson et al. (2016)

predominantly to Gs, although it can couple to other G-proteins and there is some evidence that this results in changes of the pharmacology of the receptor, depending on the RAMP (Weston et al. 2016a).

There is no crystal structure for an entire CLR/RAMP1 heterodimer, but a number of structures are available showing the ECD of this complex, in combination with both the non-peptide antagonists (Olcegepant and Telcagepant) and a modified C-terminal CGRP fragment (Table 1). There is also a structure of AM_{22-52} bound to the ECDs of CLR and RAMP2 and of an analogue of sCT_{8-32} bound to the ECD of the CTR. In addition, there are cryo-em structures for the full-length CTR and glucagon receptors (although the N-terminus of the CTR is not properly resolved) and a crystal structure of the full-length GLP-1 receptor (Liang et al. 2017; Jazayeri et al. 2017; Zhang et al. 2017). Thus, it is possible to model the TMD of CLR with some confidence, although the interaction between this part of the molecule and the juxtamembrane region of RAMP1 remains speculative. Figure 1 shows one possible structure for the complex between full-length CLR and RAMP1, but it should be considered as illustrative only. In this chapter, the structure of each component of the receptor will be considered, starting at the ECD. The human receptors will be considered, unless otherwise stated. There is well over 90% identity between the components of the human receptors and those from mammalian species normally used as models, although, particularly in RAMP1, there are some differences that are pharmacologically relevant. These will be discussed in the text.

2 The Extracellular Domain of Calcitonin Receptor-Like Receptor/Receptor Activity-Modifying Protein 1; Peptide Binding

The basic structure of the ECD of CLR/RAMP1 is shown in Fig. 2. The first 22 amino acids of CLR form a signal peptide and it is assumed that this is cleaved. The next 15 or so amino acids are not resolved in existing crystal structures but beyond this is a clear section of alpha helix; the unresolved extreme N-terminus may at least partly form a continuation of this in the full-length receptor. One side of the

Fig. 1 Speculative structure
of the full-length calcitonin
receptor-like receptor (CLR)/
receptor activity-modifying
protein 1 (RAMP1) complex
with bound CGRP. Yellow,
CLR; blue, RAMP1;
green, CGRP

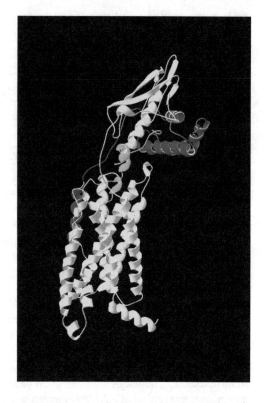

N-terminus faces the bound CGRP and the other RAMP1. Beyond the alpha helix are a three sets of beta sheets connected by loops with a final section of helix, which probably leads into the "stalk", a helix that connects TM helix 1 (TM1) with the ECD. RAMP1 consists of three helices, with helices 2 and 3 facing CLR.

The analogues of calcitonin, CGRP and AM used for crystallisation all bind in an extended form to the ECDs of CTR or CLR, unlike other class B GPCRs which have a largely alpha-helical conformation (Booe et al. 2015; Johansson et al. 2016). Examination of the structures of these bound to their cognate receptors gives some insights into the specificity of the CLR/RAMP1 complex for CGRP. The ligands all terminate in beta turns, explaining why analogues where this structure is strengthened have high affinity (reviewed in Watkins et al. 2013); however, for CGRP and AM this is a beta I-turn whereas for the calcitonin analogue it is a beta II-turn. Importantly, the C-terminal Pro of calcitonin is directed towards CTR, close to W79 and Y131. By contrast for CGRP and AM, the terminal Phe or Tyr faces towards the RAMP (Fig. 2). F37 of CGRP contacts W84 of RAMP1. In RAMP2, the equivalent of W84 is F111 which cannot contact AM. Instead, E101 contacts Y52 of AM. In RAMP1, the equivalent residue of E101 is W74 but this is not in contact with CGRP. There are no further direct contacts between either peptide and the RAMPs; instead, there are multiple contacts between the peptides and CLR or CTR as appropriate (Booe et al. 2015; Lee et al. 2016). There is evidence for some small but potentially

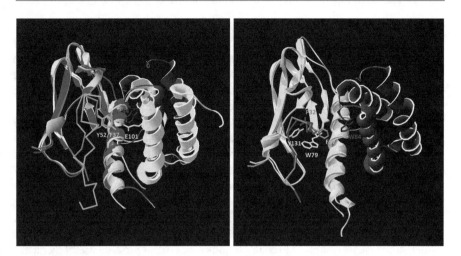

Fig. 2 Structure of the ECDs of CLR/RAMP1 and RAMP2 and calcitonin receptor (CTR). Left: CLR (pink) with RAMP1 (red) and bound $CGRP_{27-37}$ [D^{31},P^{34},F^{35}] (grey) superimposed on CLR (blue) with RAMP2 (yellow) and AM_{22-52} (green). F37 ($CGRP_{27-37}$ [D^{31},P^{34},F^{35}]), Y52 (AM_{22-52}), W84 (RAMP1) and E101 (RAMP2) are shown in line form. Right: CLR with RAMP1 and bound $CGRP_{27-37}$ [D^{31},P^{34},F^{35}] (colours as previously) superimposed on CTR (yellow) and [BrPhe^{22}]sCT$_{8-32}$ (blue). F37 ($CGRP_{27-37}$ [D^{31},P^{34},F^{35}]), P32 ([BrPhe^{22}]sCT$_{8-32}$), W84 (RAMP1 and W79/Y131 (CTR) are shown in line form

significant RAMP-dependant shifts in the conformation of the contact residues on CLR, suggesting that the RAMPs act in part by allostery (Booe et al. 2015). This effect is probably mediated via changes in the β1–β2 loop of CLR which forms part of the peptide-binding pocket (Booe et al. 2018). The detailed knowledge of how peptides interact with the C-termini of the different receptors has already resulted in the rationale design of analogues with enhanced affinity or changed selectivity (Booe et al. 2018).

Whilst the crystal structure of [BrPhe^{22}]sCT$_{8-32}$ bound to the ECD of the CTR receptor provides important insights into the selectivity of CTR for calcitonin, there remain other subtleties. There is a report that N-glycosylation of the CTR enhances affinity for salmon calcitonin (Ho et al. 1999). Studies of the isolated ECD of the CTR have shown that this is due to the presence of a single N-acetyl glucosamine attached to N130 and it enhances the binding of FITC-AC413(6–25) Y25P (an amylin analogue modified to bind to the CTR) and sCT$_{8-32}$ to the ECD of the CTR and FITC-AC413(6–25) to a fusion of the ECDs of CTR and RAMP2 (i.e. the ECD of the AMY$_2$ receptor) (Lee et al. 2017). On the basis of modelling, the authors suggest that the glycan acts allosterically to enhance ligand affinity; as N130 faces away from the bound CT analogue in the crystal structure, it is difficult to envisage any direct mechanism. It is not known if any similar mechanism exists for CLR-based receptors.

A major puzzle with amylin receptors is that C-terminal residue, Y37-amide, plays very little role in its binding; for the AM and CGRP, the C-terminus makes crucial contacts with the RAMPs. This implies that at amylin receptors, the RAMPs act largely through allosteric mechanisms (Gingell et al. 2016; Lee et al. 2017). Molecular modelling has suggested that the RAMPs might influence the dynamics of loop 5 and residues immediately C-terminal of the CTR (Gingell et al. 2016). On the basis of the structure of the salmon calcitonin analogue bound to the ECD of the CTR, it has been suggested that the RAMPs change the orientation of R126 in loop 5 of CTR to enhance the affinity of the receptor for amylin (Johansson et al. 2016). RAMPs enhance the affinity of a CTR/CLR orthologue from *Branchiostoma floridae* to bind its calcitonin/CGRP orthologues (Sekiguchi et al. 2016). The authors of this study consider that this arises from the RAMPs enhancing cell surface expression of the CTR/CLR orthologue. However, C-termini of the calcitonin/CGRP orthologues which this receptor binds appear much closer to calcitonin than CGRP (a shared terminal Pro-amide) and so it is not clear that they make direct contact with the RAMPs. If this is correct, it would further strengthen the case for an allosteric role of RAMPs. In turn, this raises the possibility that allosteric modulators acting on the ECDs of the RAMP/GPCR complexes might produce useful therapeutic effects, if suitable drug-binding pockets can be targeted.

3   The Extracellular Domain of Calcitonin Receptor-Like Receptor/Receptor Activity-Modifying Protein 1; Non-peptide Antagonist Binding

Crystal structures are available showing the interactions of ligands with CLR/RAMP1 as well as the CLR/RAMP1 structure by itself (Fig. 3). The two non-peptide antagonists of the CGRP receptor, olcegepant and telcagepant, bind to the ECD of the receptor and sit between RAMP1 and CLR in the groove that is used by the peptides for binding, making contacts with amino acids in both subunits. A set of tryptophans (W84 and W74 of RAMP1, W72 and W121 of CLR) are of particular importance in the interactions with the antagonists; W72 and W84 produce a predominantly hydrophobic pocket for the docking of the antagonists (ter Haar et al. 2010). Residue 74 is normally a basic amino acid in non-primates, explaining the species dependency of these antagonists (Mallee et al. 2002). It can readily be appreciated from these structures that the endogenous peptides will not be able to interact with the ECD with the antagonists in place. Although the affinity of olcegepant is higher than that of telcagepant, the latter is the more efficient ligand when considering the ratio of molecular weight to affinity. This arises from an additional hydrogen bond formed with T122 of CLR and the ability of the diflurophenyl group to fit further into the RAMP1-binding pocket compared to the larger dibromotyrosyl group of olcegepant (ter Haar et al. 2010).

Compared to the unliganded receptor, both antagonists cause some rearrangement of residue side chains when they bind. There is movement of R119 and a rotation of W72 of CLR as well as a slight movement of W74 of RAMP1. The rotation of W72 is of particular significance as this forms a "shelf" on which the piperidine group of the antagonists can sit (ter Haar et al. 2010). Olcegepant but not telcagepant shows pathway-selective antagonism for the AMY$_{1(a)}$ receptor; this suggests that at least at this receptor, there are further conformational changes produced when it binds to

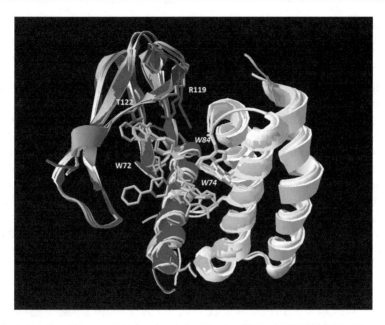

Fig. 3 Interaction of non-peptide antagonists with the ECD of CLR/RAMP1. The overlay compares CLR (grey)/RAMP1 (white) in the absence of bound ligand with CLR (red)/RAMP1 (yellow) with olcegepant (pink) and CLR (blue)/RAMP1 (green) with telcagepant (light blue). W74 and W84 are shown on the RAMP1 structures, with W72, R119 and T122 on CLR

the ECD (Walker et al. 2017). The mechanism of action of pathway-specific effects of agents that bind exclusively to the ECDs of the receptors is unclear. There is evidence for some family B GPCRs that changes in the ECD conformation are important in receptor activation; effects on the ECDs could be transmitted to the TM domain of the receptor via the extracellular loops (see below, Sect. 4). It is possible that a similar mechanism is at work with CLR- and CTR-based receptors, although the RAMP ECD–TM interfaces introduce additional complexity (but perhaps also new opportunities for drug discovery).

4   The TM Domains of Calcitonin Receptor and Calcitonin Receptor-Like Receptor

There is a mass of mutagenesis data on the ECLs and TMDs of CLR, but a clear interpretation of this remains elusive (Barwell et al. 2012; Woolley et al. 2013; Vohra et al. 2013; Watkins et al. 2016). The recent availability of a cryo-em structure of CT docked to the CTR is an important development, despite its relatively low resolution (Liang et al. 2017). A major barrier to applying this structure to the binding of CGRP to its receptors is a lack of certainty as the role of the RAMPs at this level of the receptor. Experimental evidence has been produced for a RAMP-binding interface on the receptor involving mainly TMs 6 and 7 or 1–5 (see Barwell et al. 2012 for review).

Alanine scans of the ECLs of CLR show that residues in all three are important, but ECL2 is of particular significance. Residues at the top of this loop influence the binding of $CGRP_{8-37}$, suggesting that residues 1–7 are located at the base of the loop. This is consistent with the cryo-em structure of CT bound to the CTR, where the authors suggest that S5 and T6 of CT are in contact with $His^{5.40b}$ (residue 302 in CTR, 295 in CLR), high on TM5 with N14 of CT making contacts with the backbone of ECL2 (Liang et al. 2017). Whilst it has been customary to consider agonist activity as being determined just by the disulphide-bonded loops of these receptors, it has recently been shown that $CGRP_{8-37}$ has partial agonist activity at the $AMY_{1(a)}$ receptor (Walker et al. 2017); it is tempting to suggest that this involves an interaction of residues 8 to around 14 with ECL2 or 3 of CTR. The pathway-selective antagonism of olcegepant (which binds exclusively to the ECD) at the $AMY_{1(a)}$ receptor further points to probably interactions between the ECDs and the ECL of this receptor (Walker et al. 2017). Structural data suggests that there is considerable movement of the ECD on activation of the glucagon receptor (Zhang et al. 2017); perhaps something similar happens with the $AMY_{1(a)}$ receptor.

Whilst ECL2 is the main contact for bound CT and is also the most significant loop for CGRP binding, the peptide-binding pocket is shaped by interactions with the other two ECLs. Mutagenesis has shown that the RAMPs influence the extracellular loops of CLR in their ability to determine CGRP potency (Watkins et al. 2016) and it is easy to envisage how by changing the orientation of these loops, peptide selectivity might be changed. ECL3 has also been shown to be important in AM receptors (Kuwasako et al. 2012). It is also interesting to note how mutation of a few residues at the base of ECLs can change specificity for receptors; thus, $H374A^{7.47b}$ increases AM potency 100-fold at the CGRP receptor (Woolley et al. 2017). This demonstrates that peptide selectivity is determined both by interactions at the N- and C-termini; a favourable interaction between the N-terminus of the peptide and juxtamembrane region can outweigh an unfavourable interaction at the C-terminus with the RAMP. It may be possible to design drugs that act on the TM domain of CLR but which can retain some selectivity in terms of receptor activation.

Beyond the ECLs, an extensive network of both hydrophobic and hydrophilic TM residues are important in controlling CLR activation, as judged by mutagenesis (Fig. 4). Many of these residues are also implicated in the activation of other family B GPCRs, suggesting that there are common themes to receptor activation (Wootten et al. 2013, 2016; Cordomi et al. 2017; Singh et al. 2015). Furthermore, RAMP influences rapidly diminish away from the base of the ECLs. Structure–activity studies show that Thr6 of CGRP is conserved in all members of the CGRP/AM/CT family and is essential for receptor activation (Watkins et al. 2013). It is tempting to see this interacting with $H^{5.40b}$ in both CTR and CLR. Combined with other interactions at the ECLs, this may be sufficient to cause a rearrangement of contacts at the top of the receptor, leading to the pivoting of TMs 5 and 6 around TM3. There is a pronounced proline kink in TM6, so any movement of this helix will cause the opening of a G-protein-binding pocket on the cytoplasmic face of the receptor.

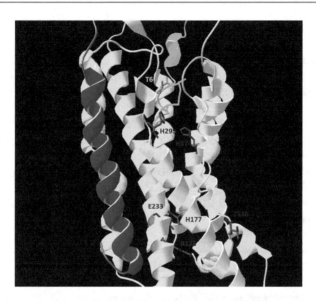

Fig. 4 Important TM residues in CLR. Yellow, CLR; blue, RAMP1 (speculative); green, CGRP. T6 on CGRP is in blue, R173, R177, E233, H295, T338 and H374 of CLR are shown in brown

This requires the rupture of a hydrogen bond between T388$^{6.42b}$ at the base of TM6 and E233$^{3.50b}$, which in turn instead interacts with H177$^{2.50b}$. R173$^{2.46b}$ drops down, allowing contact with Gs (Barwell et al. 2013). These movements are also linked to a shift in H8 and the base of TM7 (Vohra et al. 2013).

An overall model of receptor activation is that the peptide agonists at CLR- and CTR-based receptors make a variety of contacts from the top of the ECLs to the beginning of the TM domain. These cause reorientation of the ECLs, with residues either moving towards or away from the bound peptide. These changes get funnelled to the amino acids that control the orientation of TMs 3, 5 and 6 and so trigger changes in the TM bundle leading to the opening of the binding pocket for G-proteins or arrestins.

Significant work has demonstrated how RAMPs can influence ligand bias and affinity at amylin receptors (Morfis et al. 2008; Udawela et al. 2006a, b, 2008). This suggests that the C-terminus of the RAMP influences G-protein coupling. As noted above, RAMPs also change peptide G-protein-coupling preferences and hence receptor pharmacology of CLR; modelling indicates that the C-terminus of the RAMP can directly interact with the probable G-protein-binding pocket of CLR (Weston et al. 2016b). The different G-proteins themselves will be expected to act as allosteric modulators of the GPCRs, potentially changing their ligand-binding properties. This may explain why it has been reported that at the CGRP receptor, AM is more potent than CGRP at Gi coupling (Weston et al. 2016a); it would be useful to have further confirmation of this observation as it has important implications for ligand selectivity and receptor pharmacology.

5   The C-Terminus

The C-terminus of CLR remains little explored as regard its role in the CGRP receptor, although it has been studied in the AM1 receptor. Deletion of the entire C-terminus including H8 greatly impairs CLR expression, although the receptor that is able to reach the cell surface can still couple to Gs. There is a serine/threonine rich region distal to H8 and deletion of these residues impairs receptor internalisation, presumably by disrupting phosphorylation and interaction with β-arrestin (Conner et al. 2008). For the AM1 receptor, determinants influencing Gs coupling were also observed in this region and H8 (Kuwasako et al. 2006, 2010, 2011), but this was not seen in the single study done on the CGRP receptor. It is not clear if this reflects a difference between AM1 and CGRP receptors or is a consequence of the different cell lines used to express the receptors.

6   Receptor Component Protein

Receptor component protein (RCP) is a 148-amino acid, 17-kDa peripheral membrane protein. It is required for efficient coupling of the CGRP receptor to Gαs and AM acting at the AM1 receptor, as shown by knockdown of RCP expression (Evans et al. 2000). There is no information on whether it is plays a similar role in AM2 receptors. RCP appears to physically associate with the receptor, interacting with its second intracellular loop (ICL2) (Dickerson 2013). Loss of RCP does not affect the affinity of CGRP for its receptor, or significantly alter trafficking to the cell surface and so is not required in order for CLR and RAMP1 to interact. Decreases in RCP expression have been correlated with reduced sensitivity to CGRP under a number of physiological and pathological conditions (Dickerson 2013). Interestingly, a *Drosophila* class B GPCR, CG17415, also appears to interact with a homologue of RCP and human RCP can enhance coupling of this receptor to Gαs (Johnson et al. 2005). This indicates that RCP–receptor interactions appear to have co-evolved with the emergence of class B GPCRs. RCP may be an important target for allosteric modulators.

In addition to its role as part of CGRP and AM1 receptors, RCP is a component of human RNA polymerase 3 where it is known as rpc9; homologues of this protein are found in organisms as distant as yeast. In RNA polymerase III, it forms a dimer with rpc8. It may be involved in the binding of RNA transcripts as they exit the polymerase (Hu et al. 2002). Interestingly, nuclear translocation of RCP has been observed in NIH3T3 cells following challenge with CGRP, perhaps suggesting a role in nuclear signalling in addition to facilitating Gαs coupling (Sardi et al. 2014).

7   Conclusion

We currently have clear evidence for how the C-terminus of CGRP interacts with the ECD of its receptor and there are good clues as to how the N-terminus interacts with the TM domain of CLR. It is likely that we will shortly be in possession of structures of many CLR– and CTR–RAMP complexes that are relevant to the understanding of CGRP pharmacology and these will give definite information on the structure of the entire GPCR–RAMP complex. Even with this information, some significant challenges may remain. We need to understand the interactions between the ECD and TM portions of the receptor and how they work together in determining ligand binding and receptor activation. More knowledge is needed on how G-proteins may act to influence ligand binding and the roles of RCP are still poorly understood. It may be that the CLR– and CTR–RAMP complexes are particularly finely tuned allosteric machines, showing long-range interactions and that control of their dynamics is of particular significance. Knowledge of the structure of CGRP and related receptors will help rationalise the mode of action and selectivity of existing therapeutic agents and should point the way to the design of new drugs.

Acknowledgements This work was supported by the BBSRC (grant number BB/M007529/1).

References

Barwell J, Gingell JJ, Watkins HA, Archbold JK, Poyner DR, Hay DL (2012) Calcitonin and calcitonin receptor-like receptors: common themes with family B GPCRs? Br J Pharmacol 166(1):51–65

Barwell J, Wheatley M, Conner AC, Taddese B, Vohra S, Reynolds CA, Poyner DR (2013) The activation of the CGRP receptor. Biochem Soc Trans 41(1):180–184

Booe JM, Walker CS, Barwell J, Kuteyi G, Simms J, Jamaluddin MA, Warner ML, Bill RM, Harris PW, Brimble MA, Poyner DR, Hay DL, Pioszak AA (2015) Structural basis for receptor activity-modifying protein-dependent selective peptide recognition by a G protein-coupled receptor. Mol Cell 58(6):1040–1052

Booe JM, Warner ML, Roehrkasse AM, Hay DL, Pioszak AA (2018) Probing the mechanism of receptor activity-modifying protein modulation of GPCR ligand selectivity through rational design of potent adrenomedullin and calcitonin gene-related peptide antagonists. Mol Pharmacol 93:355

Conner M, Hicks MR, Dafforn T, Knowles TJ, Ludwig C, Staddon S, Overduin M, Gunther UL, Thome J, Wheatley M, Poyner DR, Conner AC (2008) Functional and biophysical analysis of the C-terminus of the CGRP-receptor; a family B GPCR. Biochemistry 47(32):8434–8444

Cordomi A, Liapakis G, Matsoukas MT (2017) Understanding Corticotropin Releasing Factor Receptor (CRFR) activation using structural models. Curr Mol Pharmacol 10(4):325–333

Dickerson IM (2013) Role of CGRP-receptor component protein (RCP) in CLR/RAMP function. Curr Protein Pept Sci 14(5):407–415

Evans BN, Rosenblatt MI, Mnayer LO, Oliver KR, Dickerson IM (2000) CGRP-RCP, a novel protein required for signal transduction at calcitonin gene-related peptide and adrenomedullin receptors. J Biol Chem 275(40):31438–31443

Gingell JJ, Simms J, Barwell J, Poyner DR, Watkins HA, Pioszak AA, Sexton PM, Hay DL (2016) An allosteric role for receptor activity-modifying proteins in defining GPCR pharmacology. Cell Discov 2:16012

Hay DL, Pioszak AA (2016) Receptor activity modifying proteins: new insights and roles. Annu Rev Pharmacol Toxicol 56:469

Hay DL, Poyner DR, Quirion R (2008) International Union of Pharmacology. LXIX. Status of the calcitonin gene-related peptide subtype 2 receptor. Pharmacol Rev 60(2):143–145

Hay DL, Garelja ML, Poyner DR, Walker CS (2017) Update on the pharmacology of calcitonin/CGRP family of peptides: IUPHAR Review 25. Br J Pharmacol 175(1):3–17

Ho HH, Gilbert MT, Nussenzveig DR, Gershengorn MC (1999) Glycosylation is important for binding to human calcitonin receptors. Biochemistry 38(6):1866–1872

Hu P, Wu S, Sun Y, Yuan CC, Kobayashi R, Myers MP, Hernandez N (2002) Characterization of human RNA polymerase III identifies orthologues for Saccharomyces cerevisiae RNA polymerase III subunits. Mol Cell Biol 22(22):8044–8055

Jazayeri A, Rappas M, Brown AJH, Kean J, Errey JC, Robertson NJ, Fiez-Vandal C, Andrews SP, Congreve M, Bortolato A, Mason JS, Baig AH, Teobald I, Dore AS, Weir M, Cooke RM, Marshall FH (2017) Crystal structure of the GLP-1 receptor bound to a peptide agonist. Nature 546(7657):254–258

Johansson E, Hansen JL, Hansen AM, Shaw AC, Becker P, Schaffer L, Reedtz-Runge S (2016) Type II turn of receptor-bound Salmon calcitonin revealed by X-ray crystallography. J Biol Chem 291(26):13689–13698

Johnson EC, Shafer OT, Trigg JS, Park J, Schooley DA, Dow JA, Taghert PH (2005) A novel diuretic hormone receptor in Drosophila: evidence for conservation of CGRP signaling. J Exp Biol 208(Pt 7):1239–1246

Kuwasako K, Cao YN, Chu CP, Iwatsubo S, Eto T, Kitamura K (2006) Functions of the cytoplasmic tails of the human receptor activity-modifying protein components of calcitonin gene-related peptide and adrenomedullin receptors. J Biol Chem 281(11):7205–7213

Kuwasako K, Kitamura K, Nagata S, Hikosaka T, Kato J (2010) Function of the cytoplasmic tail of human calcitonin receptor-like receptor in complex with receptor activity-modifying protein 2. Biochem Biophys Res Commun 392(3):380–385

Kuwasako K, Kitamura K, Nagata S, Hikosaka T, Kato J (2011) Structure-function analysis of helix 8 of human calcitonin receptor-like receptor within the adrenomedullin 1 receptor. Peptides 32(1):144–149

Kuwasako K, Hay DL, Nagata S, Hikosaka T, Kitamura K, Kato J (2012) The third extra-cellular loop of the human calcitonin receptor-like receptor is crucial for the activation of adrenomedullin signalling. Br J Pharmacol 166(1):137–150

Lee SM, Hay DL, Pioszak AA (2016) Calcitonin and amylin receptor peptide interaction mechanisms: insights into peptide-binding modes and allosteric modulation of the calcitonin receptor by receptor activity-modifying proteins. J Biol Chem 291(16):8686–8700

Lee SM, Booe JM, Gingell JJ, Sjoelund V, Hay DL, Pioszak AA (2017) N-glycosylation of asparagine 130 in the extracellular domain of the human calcitonin receptor significantly increases peptide hormone affinity. Biochemistry 56:3380

Liang YL, Khoshouei M, Radjainia M, Zhang Y, Glukhova A, Tarrasch J, Thal DM, Furness SGB, Christopoulos G, Coudrat T, Danev R, Baumeister W, Miller LJ, Christopoulos A, Kobilka BK, Wootten D, Skiniotis G, Sexton PM (2017) Phase-plate cryo-EM structure of a class B GPCR-G-protein complex. Nature 546(7656):118–123

Mallee JJ, Salvatore CA, LeBourdelles B, Oliver KR, Longmore J, Koblan KS, Kane SA (2002) Receptor activity-modifying protein 1 determines the species selectivity of non-peptide CGRP receptor antagonists. J Biol Chem 277(16):14294–14298

McLatchie LM, Fraser NJ, Main MJ, Wise A, Brown J, Thompson N, Solari R, Lee MG, Foord SM (1998) RAMPs regulate the transport and ligand specificity of the calcitonin-receptor-like receptor. Nature 393(6683):333–339

Morfis M, Tilakaratne N, Furness SG, Christopoulos G, Werry TD, Christopoulos A, Sexton PM (2008) Receptor activity-modifying proteins differentially modulate the G protein-coupling efficiency of amylin receptors. Endocrinology 149(11):5423–5431

Sardi C, Zambusi L, Finardi A, Ruffini F, Tolun AA, Dickerson IM, Righi M, Zacchetti D, Grohovaz F, Provini L, Furlan R, Morara S (2014) Involvement of calcitonin gene-related peptide and receptor component protein in experimental autoimmune encephalomyelitis. J Neuroimmunol 271(1–2):18–29

Sekiguchi T, Kuwasako K, Ogasawara M, Takahashi H, Matsubara S, Osugi T, Muramatsu I, Sasayama Y, Suzuki N, Satake H (2016) Evidence for conservation of the calcitonin superfamily and activity-regulating mechanisms in the basal chordate Branchiostoma floridae: insights into the molecular and functional evolution in chordates. J Biol Chem 291(5):2345–2356

Singh R, Ahalawat N, Murarka RK (2015) Activation of corticotropin-releasing factor 1 receptor: insights from molecular dynamics simulations. J Phys Chem B 119(7):2806–2817

ter Haar E, Koth CM, Abdul-Manan N, Swenson L, Coll JT, Lippke JA, Lepre CA, Garcia-Guzman M, Moore JM (2010) Crystal structure of the ectodomain complex of the CGRP receptor, a class-B GPCR, reveals the site of drug antagonism. Structure 18(9):1083–1093

Udawela M, Christopoulos G, Morfis M, Christopoulos A, Ye S, Tilakaratne N, Sexton PM (2006a) A critical role for the short intracellular C terminus in receptor activity-modifying protein function. Mol Pharmacol 70(5):1750–1760

Udawela M, Christopoulos G, Tilakaratne N, Christopoulos A, Albiston A, Sexton PM (2006b) Distinct receptor activity-modifying protein domains differentially modulate interaction with calcitonin receptors. Mol Pharmacol 69(6):1984–1989

Udawela M, Christopoulos G, Morfis M, Tilakaratne N, Christopoulos A, Sexton PM (2008) The effects of C-terminal truncation of receptor activity modifying proteins on the induction of amylin receptor phenotype from human CTb receptors. Regul Pept 145(1–3):65–71

Vohra S, Taddese B, Conner AC, Poyner DR, Hay DL, Barwell J, Reeves PJ, Upton GJ, Reynolds CA (2013) Similarity between class A and class B G-protein-coupled receptors exemplified through calcitonin gene-related peptide receptor modelling and mutagenesis studies. J R Soc Interface 10(79):20120846

Walker CS, Eftekhari S, Bower RL, Wilderman A, Insel PA, Edvinsson L, Waldvogel HJ, Jamaluddin MA, Russo AF, Hay DL (2015) A second trigeminal CGRP receptor: function and expression of the AMY1 receptor. Ann Clin Transl Neurol 2(6):595–608

Walker CS, Raddant AC, Woolley MJ, Russo AF, Hay DL (2017) CGRP receptor antagonist activity of olcegepant depends on the signalling pathway measured. Cephalalgia. https://doi.org/10.1177/0333102417691762

Watkins HA, Rathbone DL, Barwell J, Hay DL, Poyner DR (2013) Structure-activity relationships for alpha-calcitonin gene-related peptide. Br J Pharmacol 170(7):1308–1322

Watkins HA, Chakravarthy M, Abhayawardana RS, Gingell JJ, Garelja M, Pardamwar M, McElhinney JM, Lathbridge A, Constantine A, Harris PW, Yuen TY, Brimble MA, Barwell J, Poyner DR, Woolley MJ, Conner AC, Pioszak AA, Reynolds CA, Hay DL (2016) Receptor activity-modifying proteins 2 and 3 generate adrenomedullin receptor subtypes with distinct molecular properties. J Biol Chem 291(22):11657–11675

Weston C, Winfield I, Harris M, Hodgson R, Shah A, Dowell SJ, Mobarec JC, Woodcock DA, Reynolds CA, Poyner DR, Watkins HA, Ladds G (2016a) Receptor activity-modifying protein-directed G protein signaling specificity for the calcitonin gene-related peptide family of receptors. J Biol Chem 291(49):25763

Weston C, Winfield I, Harris M, Hodgson R, Shah A, Dowell SJ, Mobarec JC, Woodlock DA, Reynolds CA, Poyner DR, Watkins HA, Ladds G (2016b) Receptor activity-modifying protein-directed G protein signaling specificity for the calcitonin gene-related peptide family of receptors. J Biol Chem 291(42):21925–21944

Woolley MJ, Watkins HA, Taddese B, Karakullukcu ZG, Barwell J, Smith KJ, Hay DL, Poyner DR, Reynolds CA, Conner AC (2013) The role of ECL2 in CGRP receptor activation: a combined modelling and experimental approach. J R Soc Interface 10(88):20130589

Woolley MJ, Reynolds CA, Simms J, Walker CS, Mobarec JC, Garelja ML, Conner AC, Poyner DR, Hay DL (2017) Receptor activity-modifying protein dependent and independent activation mechanisms in the coupling of calcitonin gene-related peptide and adrenomedullin receptors to Gs. Biochem Pharmacol 142:96–110

Wootten D, Simms J, Miller LJ, Christopoulos A, Sexton PM (2013) Polar transmembrane interactions drive formation of ligand-specific and signal pathway-biased family B G protein-coupled receptor conformations. Proc Natl Acad Sci U S A 110:5211

Wootten D, Reynolds CA, Koole C, Smith KJ, Mobarec JC, Simms J, Quon T, Coudrat T, Furness SG, Miller LJ, Christopoulos A, Sexton PM (2016) A hydrogen-bonded polar network in the core of the glucagon-like peptide-1 receptor is a fulcrum for biased agonism: lessons from class B crystal structures. Mol Pharmacol 89(3):335–347

Zhang H, Qiao A, Yang D, Yang L, Dai A, de Graaf C, Reedtz-Runge S, Dharmarajan V, Zhang H, Han GW, Grant TD, Sierra RG, Weierstall U, Nelson G, Liu W, Wu Y, Ma L, Cai X, Lin G, Wu X, Geng Z, Dong Y, Song G, Griffin PR, Lau J, Cherezov V, Yang H, Hanson MA, Stevens RC, Zhao Q, Jiang H, Wang MW, Wu B (2017) Structure of the full-length glucagon class B G-protein-coupled receptor. Nature 546(7657):259–264

CGRP Receptor Signalling Pathways

Graeme S. Cottrell

Contents

Abstract

Calcitonin gene-related peptide (CGRP) is a promiscuous peptide, similar to many other members of the calcitonin family of peptides. The potential of CGRP to act on many different receptors with differing affinities and efficacies makes deciphering the signalling from the CGRP receptor a challenging task for researchers.

Although it is not a typical G protein-coupled receptor (GPCR), in that it is composed not just of a GPCR, the CGRP receptor activates many of the same signalling pathways common for other GPCRs. This includes the family of G proteins and a variety of protein kinases and transcription factors. It is now also

G. S. Cottrell (✉)
University of Reading, Reading, UK
e-mail: g.s.cottrell@reading.ac.uk

© Springer International Publishing AG, part of Springer Nature 2018   37
S. D. Brain, P. Geppetti (eds.), *Calcitonin Gene-Related Peptide (CGRP) Mechanisms*,
Handbook of Experimental Pharmacology 255, https://doi.org/10.1007/164_2018_130

clear that in addition to the initiation of cell-surface signalling, GPCRs, including the CGRP receptor, also activate distinct signalling pathways as the receptor is trafficking along the endocytic conduit.

Given CGRP's characteristic of activating multiple GPCRs, we will first consider the complex of calcitonin receptor-like receptor (CLR) and receptor activity-modifying protein 1 (RAMP1) as the CGRP receptor. We will discuss the discovery of the CGRP receptor components, the molecular mechanisms controlling its internalization and post-endocytic trafficking (recycling and degradation) and the diverse signalling cascades that are elicited by this receptor in model cell lines. We will then discuss CGRP-mediated signalling pathways in primary cells pertinent to migraine including neurons, glial cells and vascular smooth muscle cells.

Investigation of all the CGRP- and CGRP receptor-mediated signalling cascades is vital if we are to fully understand CGRP's role in migraine and will no doubt unearth new targets for the treatment of migraine and other CGRP-driven diseases.

Keywords

Calcitonin gene-related peptide · Calcitonin receptor-like receptor · G protein · Protein kinase · Receptor activity-modifying protein · Signalling · Trafficking

Abbreviations

ATP	Adenosine triphosphate
cAMP	Cyclic adenosine monophosphate
CGRP	Calcitonin gene-related peptide
CLR	Calcitonin receptor-like receptor
ECE1	Endothelin-converting enzyme 1
ERK	Extracellular-regulated protein kinase
ET_A	Endothelin A receptor
GPCR	G protein-coupled receptor
IL	Interleukin
JNK	c-Jun N-terminal kinase
PKA	Protein kinase A
PKC	Protein kinase C
NO	Nitric oxide
NOS	Nitric oxide synthase
RAMP	Receptor activity-modifying protein
RCP	Receptor component protein

1   The Discovery of the Calcitonin Gene-Related Peptide Receptor

Although Amara et al. (1982) discovered calcitonin gene-related peptide (CGRP) in 1982, it was not until many years later that the identity of the receptor for CGRP was confirmed (McLatchie et al. 1998). A crucial development in the discovery of the CGRP receptor was the identification of a new family of single transmembrane proteins called receptor activity-modifying proteins (RAMPs). It had long been suspected that the G protein-coupled receptor (GPCR), calcitonin receptor-like receptor (CLR) was the receptor for CGRP. However, when this protein was expressed in model cell lines, CLR did not traffic to the cell-surface and CGRP was unable to elicit cellular responses typical of CGRP.

The human neuroblastoma cell line, SK-N-MC, was well-known to bind radio-labelled CGRP and CGRP promoted signalling following incubation with CGRP (Van Valen et al. 1990). McLatchie et al. (1998) used an expression cloning strategy in *Xenopus* oocytes that relied on a CGRP-mediated signalling pathway to activate a co-injected cystic fibrosis transmembrane regulator as a physiological read-out. Repeated subdivision of a positive pool of cDNA clones led to the isolation of a gene encoding a 148-amino acid protein that the authors named RAMP1. Initially, the authors were surprised as they expected the CGRP receptor to be a GPCR. However, further experimentation revealed that co-expression of RAMP1 with CLR was required in order to yield a high affinity CGRP receptor.

Although it was initially proposed that the role of RAMP1 was only to transport the CGRP receptor (CLR) to the cell-surface (McLatchie et al. 1998), it is now known that both CLR and RAMP1 play a vital role on the recognition of CGRP (Banerjee et al. 2006; Kuwasako et al. 2003). Indeed, RAMP1 fulfils multiple roles in the formation of a mature CGRP receptor. RAMP1 co-expression promotes the trafficking of CLR (together with RAMP1) to the cell-surface, increases the molecular mass of CLR by promoting the glycosylation process, actively participates in the binding of CGRP at the cell-surface (Fraser et al. 1999; McLatchie et al. 1998) and plays an important role in the post-endocytic sorting of CLR [reviewed in Klein et al. (2016)].

Since the discovery of RAMP1, we now know that this protein family has two other members, that also heterodimerize with CLR, forming receptors for adrenomedullin (also a member of the calcitonin family of peptides) (McLatchie et al. 1998). In addition, RAMPs also modify other GPCRs to generate receptors for other members of the calcitonin family of peptides [reviewed in Hay and Pioszak (2016)].

2   Calcitonin Gene-Related Peptide Receptors Mediate G Protein-Dependent Signalling

The first report of a signalling pathway evoked by CGRP was the cAMP-dependent effect of CGRP on amylase release from guinea pig pancreatic acinar cells (Seifert et al. 1985). Further evidence of the involvement of adenylate cyclase in signalling by CGRP was subsequently observed in cat cerebral arteries (Edvinsson et al. 1985), rat striated muscle (Kobayashi et al. 1987), chick skeletal muscle (Laufer and Changeux 1987) and mouse diaphragm (Takami et al. 1986). The first discovery that CGRP signalled through G proteins was reported by Takamori and Yoshikawa (1989). The experimental evidence was obtained by measuring twitch force in curarized rat skeletal muscle and this early study provided evidence that CGRP-mediated signalling occurred through $G\alpha_s$, but not through $G\alpha_i$ (Takamori and Yoshikawa 1989). Cholera toxin enhanced the effect of CGRP on twitch, whereas pertussis toxin had no effect (Takamori and Yoshikawa 1989).

It is now well established that the CGRP receptor couples to G proteins to initiate signalling (Fig. 1). During the cloning of the RAMPs, signalling through the CGRP receptor was examined by recording increases in intracellular cAMP (McLatchie et al. 1998). Thus, CGRP receptors couple to $G\alpha_s$-type G proteins to stimulate the activity of cell-surface enzyme adenylate cyclase, which subsequently converts adenosine triphosphate (ATP) to cyclic adenosine monophosphate (cAMP). Typically, increases in cAMP then stimulate the activation of protein kinase A (PKA). Another G protein-mediated effect is the mobilization of intracellular calcium from the endoplasmic reticulum. Activation of $G\alpha_{q/11}$ proteins promotes the activity of phospholipase Cβ, which cleaves the cell-surface located lipid, phosphatidylinositol 4,5-bisphosphate, leaving diacylglycerol anchored to the cell-surface and inositol 1,4,5-trisphosphate free to activate inositol 1,4,5-trisphosphate receptors present on the endoplasmic reticulum, which in turn causes an efflux of Ca^{2+} into the cytoplasm. Diacylglycerol promotes the activity of protein kinase C (PKC). Although, the CGRP receptor is mainly considered a $G\alpha_s$-coupled GPCR, it is becoming clear that it is not as simple as a GPCR being either $G\alpha_s$-, $G\alpha_i$- or $G\alpha_{q/11}$-coupled. After the cloning of RAMPs, the same laboratory investigated the G protein coupling of CGRP receptors in Swiss 3T3 cells and *Xenopus* oocytes and reported coupling to both pertussis toxin-sensitive and toxin-insensitive G proteins, inferring that the CGRP receptor can couple to both $G\alpha_s$ and $G\alpha_i$ proteins (Main et al. 1998). However, another interpretation could be that the pertussis toxin affected the balance of the stimulatory and inhibitory processes responsible for cAMP production.

The first evidence that CGRP promotes coupling of CLR to $G\alpha_{q/11}$ proteins was observed in 293 cells (Aiyar et al. 1999). In agreement with this and using a fura-2/AM-based Ca^{2+}-assay, another group reported increases in intracellular calcium levels following exposure of 293 cells expressing CLR and RAMP1 to CGRP (Kuwasako et al. 2000). In the study by Aiyar et al. (1999), the calcium released in response to a CGRP challenge was derived from a thapsigargin-sensitive store, indicating an endoplasmic reticulum-dependent release. Further studies examining the post-endocytic sorting and signalling of the CGRP receptor in 293 cells also confirmed that CGRP promotes the mobilization of intracellular calcium (Cottrell et al. 2007; Padilla et al. 2007).

The issue of CGRP receptor G protein coupling, as with many GPCRs, is now much more complex than first thought. A study using both yeast and mammalian cells (293 cells) demonstrated that CGRP receptors couple to $G\alpha_s$-, $G\alpha_i$- and $G\alpha_{q/11}$

A. CGRP receptor Gα_s signalling

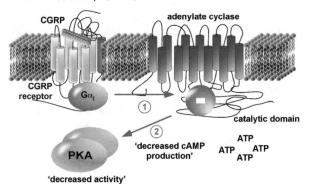

B. CGRP receptor Gα_i signalling

C. CGRP receptor Gα_q/11 signalling

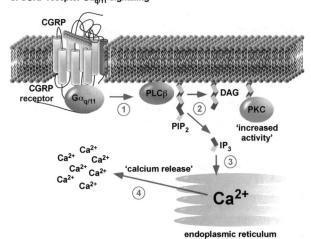

Fig. 1 Cell-surface Gα protein signalling from CGRP receptors. Following of exposure of CGRP receptors to CGRP, CGRP receptors couple to (**a**) Gα_s proteins, (**b**) Gα_i proteins and (**c**) Gα_q/11 proteins. (**a, step 1**) Gα_s proteins promote the activity of adenylate cyclase, which converts ATP to

proteins (Weston et al. 2016), with the resultant signalling dependent on the relative expression levels of the G proteins. This study highlights the importance of choosing an appropriate model system to study CGRP-induced signalling pathways.

3 Calcitonin Gene-Related Peptide Receptor Component Protein

Before the cloning of RAMPs (McLatchie et al. 1998), a cytosolic protein named receptor component protein (RCP) that conferred CGRP responsiveness in *Xenopus* oocytes was identified by expression cloning using a cochlear hair cell cDNA library, using activation of the cystic fibrosis transmembrane conductance regulator as a signalling read-out (Luebke et al. 1996). RCP is a 16-kDa protein that lacks any predicted membrane spanning domains that would explain how it could be a receptor for CGRP; the authors surmised that it potentiated signals generated by an endogenous CGRP receptor in the oocytes. Supporting this hypothesis, both CLR and RAMP1 have subsequently been identified in *Xenopus* oocytes (Guillemare et al. 1994; Klein et al. 2002; Kline et al. 1988).

After the cloning of CLR and RAMP1, it was thought that these two proteins alone were sufficient to form a high affinity CGRP receptor. However, much of the work examining CGRP receptor signalling was performed in cell lines that endogenously expressed RCP. A study further examining the role of RCP in CGRP-induced signalling identified a major confounding issue when examining its role on CGRP-mediated signalling. All the model cell lines they screened, including many cell lines previously used in studies on CGRP receptor, contained endogenous RCP (Evans et al. 2000). In order to study the role of RCP in CGRP receptor signalling, the authors silenced RCP expression using RCP antisense RNA, confirming much reduced expression by western blotting. Although the levels of CGRP receptor present at the cell-surface were unaffected by RCP knockdown, as confirmed by ligand binding, CGRP-mediated cAMP production was a third of that compared to the same cells expressing RCP (Evans et al. 2000). Thus, indicating an important role for RCP in CGRP-induced cAMP production.

It is now clear that CLR, RAMP1 and RCP form a complex at the cell-surface yielding a high affinity CGRP receptor. RCP co-immunoprecipitates with CLR (Evans et al. 2000) and RAMP1 (Prado et al. 2001). Furthermore, it is now clear

Fig. 1 (continued) cyclic AMP (cAMP, a second messenger) promoting accumulation in the cell. (**a, step 2**) cAMP binds to the regulatory subunit of protein kinase A (PKA) stimulating its kinase activity. (**b, step 1**) $G\alpha_i$ proteins inhibit the activity of the catalytic domain of adenylate cyclase, reducing the levels of cyclic AMP in the cell. (**b, step 2**) Thus, there is less cAMP available to bind to the regulatory subunit of PKA and its activity is diminished. (**C, step 1**) $G\alpha_{q/11}$ proteins increase the activity of phospholipase Cβ which (**c, step 2**) processes phosphatidylinositol 4,5-bisphosphate leaving diacylglycerol anchored to the cell-surface which recruits protein kinase C (PKC). (**c, step 3**) The released inositol 1,4,5-trisphosphate (IP$_3$) diffuses and promotes the opening of inositol trisphosphate receptors on the endoplasmic reticulum which act as calcium channels and (**c, step 4**) intracellular levels of calcium ions increase

that RCP interacts with CLR via its second intracellular cytoplasmic loop (Egea and Dickerson 2012). The interaction between RCP and the second intracellular cytoplasmic loop was first confirmed using a yeast two-hybrid system. No such interactions using the other cytoplasmic loops or the C-terminal tail were observed. Co-immunoprecipitation experiments confirmed the interaction or lack of interaction between these intracellular portions of CLR and RCP (Egea and Dickerson 2012). Furthermore, expression of the second intracellular cytoplasmic loop of CLR acted as a dominant-negative for CGRP-induced signalling, reducing cAMP production by 74%, without significantly affecting the EC_{50} (Egea and Dickerson 2012). Interestingly, the C-terminal tail of CLR was also inhibitory towards CGRP-induced signalling, suggesting a role for this portion of the GPCR in activating cAMP production. The loss of maximal effect on cAMP production without loss of efficacy at the CGRP receptor could have been explained in two ways, by a reduction in the number of CGRP receptors at the cell-surface or a loss of signalling from a constant number of cell-surface CGRP receptors. Radioligand binding and ELISA experiments determined that expression of the second intracellular cytoplasmic loop of CLR had no effect on the trafficking of CLR and RAMP1 to the cell-surface, indicating that the loss of the RCP–CLR interaction accounted for the reduction in CGRP-induced cAMP accumulation (Egea and Dickerson 2012). Thus, confirming a role for RCP in CGRP receptor signalling either by direct coupling of the membrane traversing components of the CGRP receptor to intracellular signalling proteins (G proteins) or by altering the microdomain localization of CLR and RAMP1 at the cell surface. The exact mechanism by which RCP exerts its effects on CGRP receptor-mediated cell signalling remains unknown.

4   Calcitonin Gene-Related Peptide Receptor Internalization and Trafficking

Stimulation of most GPCRs promotes their removal from the cell-surface to intracellular compartments and the CGRP receptor is no exception (Kuwasako et al. 2000) (Fig. 2). A green fluorescent protein-tagged CLR was used to visualize the internalization of the CGRP receptor after exposure to CGRP (Kuwasako et al. 2000). Typically, upon activation GPCRs undergo a conformational change that leads to the phosphorylation of serine and threonine residues on the receptor by protein kinases belonging to the G protein-coupled receptor kinase (GRK) family. Using 293 cells metabolically labelled with $^{32}P[P_i]$, it was discovered that CLR, but not RAMP1, was phosphorylated within 5 min after exposure to CGRP (Hilairet et al. 2001). Although the GRKs responsible for CGRP receptor phosphorylation have yet to be definitively identified, a study by Aiyar et al. proposed the involvement of GRK6 (Aiyar et al. 2000). However, it is unclear from this study whether CLR was co-expressed with RAMP1 and as such, care must be taken when considering the validity of the observations. The post-translational phosphorylation increases the affinity of the CGRP receptor for cytosolic proteins called β-arrestins

CGRP receptor trafficking and endosomal signalling

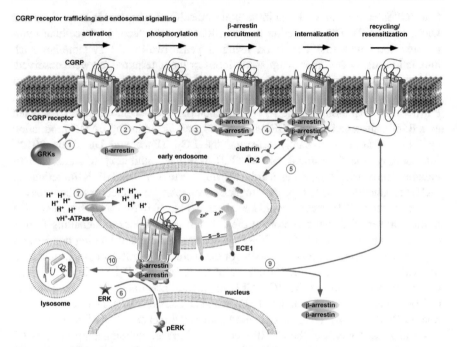

Fig. 2 Trafficking, sorting and endosomal signalling of CGRP receptors. (1) CGRP binds to the CGRP receptor promoting a conformational change and G protein-coupled receptor kinases (GRKs) phosphorylate the CGRP receptor. (2) The phosphorylated CGRP receptor has an increased affinity for β-arrestins and (3) β-arrestins translocate from the cytosol to interact with the CGRP receptor at the cell-surface. (4) β-arrestins act as a molecular scaffold recruiting clathrin and AP-2 to (5) facilitate internalization to early endosomes. (6) In the early endosomes, activated CGRP receptors promote the phosphorylation of extracellular-regulated protein kinase (ERK), which then translocates to the nucleus. (7) Vacuolar H^+-ATPase (vH^+-ATPase) pumps in protons acidifying the vesicle and (8) the changing pH reduces the affinity of CGRP for its receptor, CGRP dissociates and is degraded by endothelin-converting enzyme 1 (ECE1). (9) β-arrestins are released from the CGRP receptor which is presumably dephosphorylated and then trafficked back the cell-surface (recycling) to mediate resensitization. (10) Alternatively, if exposed to CCGRP for sustained periods, the CGRP receptor is trafficked to lysosomes for degradation

which translocate to the cell-surface to interact with the CGRP receptor (Hilairet et al. 2001). It is well established that β-arrestins perform at least three conserved functions on GPCRs [reviewed in Gurevich and Gurevich (2015), Jean-Charles et al. (2017) and Peterson and Luttrell (2017)]. First, they uncouple GPCRs from G protein via steric hindrance to terminate cell-surface G protein-dependent signalling. Second, β-arrestins act as a molecular scaffold recruiting proteins such as clathrin and adapter proteins (e.g. AP-1), that are essential for the internalization of the GPCR in a process termed clathrin-dependent endocytosis. Finally, many GPCRs remain bound to β-arrestins for long periods in intracellular vesicles called endosomes (Oakley et al. 2000). In endosomes, β-arrestins act as a hub for the recruitment of signalling molecules to initiate a wave of signalling that is distinct from that initiated at the cell-surface. Internalization of the CGRP receptor is prevented by incubation of cells in hypertonic medium, indicative of a clathrin-mediated process (Kuwasako et al. 2000). A subsequent study showed that overexpression of a dominant-negative form of β-arrestin prevented internalization of CGRP receptors and that a yellow fluorescent protein-tagged β-arrestin translocated to the cell-surface after exposure to CGRP, indicating the dependence of CGRP receptor internalization on β-arrestins (Hilairet et al. 2001). Furthermore, a GTPase-defective mutant of the protein dynamin (Herskovits et al. 1993) also prevented CGRP receptor internalization (Hilairet et al. 2001). Both dynamin and β-arrestins participate in clathrin-coated vesicle-mediated endocytosis of GPCRs (Zhang et al. 1996).

Following sequestration and internalization, many GPCRs are trafficked to intracellular vesicles called endosomes. Activated CGRP receptors traffic together with CGRP to endosomes positive for early endosome-associated protein 1 (Cottrell et al. 2005, 2007; Padilla et al. 2007). The fate of the activated CGRP receptors is dependent on the duration of this exposure to CGRP (Cottrell et al. 2007). If cells expressing CGRP receptors are transiently exposed to CGRP, then CGRP receptors are efficiently recycled back to the cell-surface to mediate resensitization. In contrast, chronic exposure to CGRP shunts the receptors to a degradative pathway and the CGRP receptors are broken down by peptidases in lysosomes (Cottrell et al. 2007). The recycling of CGRP receptors is regulated by acidification of the endosomal system by H^+-ATPase and by the proteolytic breakdown of CGRP by an endosomal-located peptidase and can be prevented by inhibitors of both processes (Padilla et al. 2007). In a mechanism similar to the regulation of bioactive peptides at the cell-surface, whereby peptidases regulate the functional availability of peptides such as angiotensin II and substance P by proteolysis, endothelin-converting enzyme 1 (ECE1) regulates the availability of CGRP within endosomes. As CGRP receptors pass through the endosomal system and the vesicles mature, protons are pumped into the vesicles lowering the pH. ECE1 is inactive towards CGRP at neutral pH; however, ECE1 has an optimal pH for the proteolytic cleavage of CGRP of pH 5–5.5 (Fahnoe et al. 2000; Padilla et al. 2007). As the pH is lowered, CGRP has a lower affinity for the CGRP receptor and becomes the substrate for ECE1 in the endosome. Cleaved by ECE1, CGRP can no longer bind to the CGRP receptor. The unbinding of CGRP presumably results in a conformational change of the CGRP receptor and β-arrestins bound to the receptor in endosomes are released, translocating back to the cytosol. The CGRP receptor, freed from β-arrestins, is then presumably dephosphorylated and free to recycle back to the cell-surface in a Rab4- and Rab11-dependent mechanism (Cottrell et al. 2007; Padilla et al. 2007). This ECE1-dependent mechanism has also been shown to regulate CGRP receptors in intact arteries (McNeish et al. 2012). However, the mechanism by which a cell switches from recycling to lysosomal degradation of CGRP receptors in the presence of continued exposure to CGRP remains undetermined.

There is also the question of the molecular mechanism that targets CGRP receptors for degradation following exposure to CGRP. Degradation of receptors is an important mechanism by which cells can control the responsiveness of a cell to a particular biological signal. Degradation of receptors allows a cell to permanently reduce its ability to respond to a particular stimulus, at least until new receptors are synthesized and trafficked to the cell-surface.

Ubiquitin is a 76-amino acid protein that is covalently attached prevalently to lysine residues (Hershko et al. 1980; Schlesinger et al. 1975; Wilkinson 2005). This post-translational modification, termed ubiquitination, serves as a signal to the cell to breakdown the protein. Ubiquitin molecules themselves are efficiently cleaved off intact and recycled. The majority of cytosolic proteins are degraded by the proteolytic activity of the proteasome, large multi-protein complex with threonine residues at the heart of its active site [reviewed in Budenholzer et al. (2017)]. In contrast, membrane proteins such as GPCRs are typically degraded by peptidases in lysosomes. As described above, the CGRP receptor is no exception and in the continued presence of CGRP, both CLR and RAMP1 are degraded in lysosomes (Cottrell et al. 2007). The first study examining the mechanism that controls the targeting of a mammalian GPCR to lysosomes involved the β_2-adrenoceptor (Shenoy et al. 2001). Following agonist-induced internalization, the β_2-adrenoceptor was ubiquitinated on lysine residues and trafficked to lysosomes. Mutation of all the intracellular facing lysine residues to arginine (which cannot be ubiquitinated) resulted in a GPCR that was not ubiquitinated and instead of trafficking to lysosomes, recycled back to the cell-surface, indicating that the agonist-induced ubiquitination served as a signal to the cell to traffic the GPCR to lysosomes. This mechanism also serves as a target for other GPCRs (Jacob et al. 2005; Marchese and Benovic 2001). However, ubiquitination of GPCRs can also regulate different phases of trafficking, affecting internalization and the rate of degradation. Furthermore, not all GPCRs are ubiquitinated following agonist-induced activation. The CGRP receptor is not ubiquitinated following stimulation with CGRP, yet still traffics to lysosomes (Cottrell et al. 2007). Thus, a different molecular mechanism must exist that indicates to the cell that CLR and RAMP1 must be targeted to lysosomes. After internalization to early endosomes, receptor cargo is sorted within the multivesicular body, with multiple 'sorting proteins' with potential roles in deciding the fate of the receptor [reviewed in Szymanska et al. (2018)]. Hepatocyte growth factor-regulated tyrosine kinase substrate (HRS) is a component of multiprotein complex found on endosomal membranes (Bache et al. 2003; Komada et al. 1997). HRS is a multi-domain protein and contains a ubiquitin-interacting motif that has an essential role in endosomal sorting processes (Shih et al. 2002; Urbe et al. 2003). HRS plays a role in the trafficking of many GPCRs and indeed, the CGRP receptor (Hanyaloglu et al. 2005; Hasdemir et al. 2007; Hislop et al. 2004). However, as the CGRP receptor is not ubiquitinated, it is unlikely that this is through a direct interaction with the HRS ubiquitin-interacting motif. The exact mechanisms by which CGRP receptors are sorted and targeted to lysosomes are still unknown and warrant further investigation.

5   Calcitonin Gene-Related Peptide Receptor Signalling Pathways in Model Cell Lines

5.1   Activation of Calcitonin Gene-Related Peptide Receptors Promotes Activation of Protein Kinases

In common with many GPCRs, the CGRP receptor activates many different protein kinase cascades. From the work in model cell lines, it is well established that activated CGRP receptors can couple to $G\alpha_s$, $G\alpha_i$ or $G\alpha_{q/11}$ proteins, with the overall resulting effect dependent upon the cell type used to study the mechanism (Weston et al. 2016). If we take these results at face value, the balance between $G\alpha_s$ and $G\alpha_i$ will lead to accumulation or reduction in intracellular levels of cAMP. Alterations in cAMP levels will have a direct effect on cAMP-dependent protein kinase, commonly known as PKA. Composed of four subunits, PKA comprises two regulatory subunits and two catalytic subunits. When cAMP levels become elevated, cAMP binds to the regulatory subunits, promoting conformational change between the regulatory and catalytic subunits, unleashing the catalytic activity of PKA. Considering the $G\alpha_{q/11}$-dependent pathway, two signalling molecules in the form of diacylglycerol and inositol 1,4,5-trisphosphate are generated from the phospholipase Cβ-dependent cleavage of phosphatidylinositol 4,5-bisphosphate at the cell-surface. Diacylglycerol serves to both activate and anchor PKC to the inner leaflet of the plasma membrane. Inositol 1,4,5-trisphosphate diffuses to the endoplasmic reticulum activating inositol 1,4,5-trisphosphate receptors which open and release calcium in the cytoplasm.

As CLR and RAMP1 are commonly expressed with RAMP2, RAMP3 and calcitonin receptors and CGRP has been observed to activate adrenomedullin and amylin receptors, it is often difficult to ascertain which signalling cascades are activated by direct interaction of CGRP with CGRP receptors. In addition to this caveat, it is also clear that CGRP receptors may also couple to intracellular signalling cascades in a cell type-specific fashion. For this reason, we will first consider what is known about the protein kinases activated by CGRP receptors in model cell lines, before examining what should be considered CGRP-induced, not CGRP receptor-induced signalling pathways.

The first reported study analysing the protein kinase pathways elicited by the CGRP receptor showed that activation of the porcine CGRP receptor coupled to extracellular-regulated protein kinase (ERK) and p38 with no significant effect on the c-jun N-terminal kinase (JNK) pathway (Parameswaran et al. 2000). Both the ERK and p38 activities were dependent on time and the concentration of CGRP. Furthermore, the activation of these pathways was decreased by preincubation with the CGRP receptor antagonist, CGRP$_{8-37}$ and the PKA inhibitor, H-89 (N-[2-(p-bromocinnamylamino)ethyl]-5-isoquinolinesulfonamide dihydrochloride) (Parameswaran et al. 2000). In contrast, wortmannin, an inhibitor of phosphatidylinositol 3-kinase, only attenuated ERK activation.

The mouse CGRP receptor components CLR and RAMP1 were cloned, expressed and the CGRP receptor characterized in COS-7 cells (Miyauchi et al. 2002). The study confirmed CGRP-induced cAMP accumulation and examined subsequent kinase activation using a variety of reporter assays (Miyauchi et al. 2002). The results from this study indicated that activation of the CGRP receptor promoted the activities of PKA and ERK but had no effect on JNK or nuclear factor-kappaB signalling (Miyauchi et al. 2002). This early work has subsequently been confirmed using the human CGRP receptor in COS-7 cells (Walker et al. 2017). Here, the authors examined CGRP-mediated activation of ERK, p38 and PKA and observed PKA and ERK activation, but in contrast to the earlier study (Parameswaran et al. 2000), no CGRP receptor-dependent activation of the stress-regulated protein kinase, p38, was reported (Walker et al. 2017).

5.2   Calcitonin Gene-Related Peptide Receptor Activation of Endosomal Signalling Pathways

For a long time, it was considered that the primary function of GPCRs was to signal from the cell surface via G proteins. In particular, activation of heterotrimeric G proteins promotes the production of second messengers such as cAMP and the release of intracellular calcium. However, it is now clear that following internalization to early endosomes, β-arrestins act as molecular hubs, recruiting signalling molecules to GPCRs in endosomes promoting a second wave of GPCR signalling that is distinct from cell-surface initiated signalling [reviewed in Eichel and von Zastrow (2018), Irannejad and von Zastrow (2014) and Sposini and Hanyaloglu (2017)]. This recruitment of signalling molecules to endosomes allows GPCR to generate compartmentalized signals that still remain largely unexplored. The first report of an endosomal signalling complex was for the β$_2$-adrenoceptor (Luttrell et al. 1999). Subsequently, DeFea et al. (2000) highlighted that β-arrestin-dependent signalling from endosomal neurokinin 1 receptors regulated the proliferative and anti-apoptotic effects of substance P. Since then, a number of studies have shown the importance of endosomes as signalling platforms for pain (Cottrell et al. 2009; Jensen et al. 2017; Yarwood et al. 2017). It was clear from studies examining the trafficking of the CGRP receptor that β-arrestins are recruited by this receptor and remain associated with the receptor for long periods in endosomes (Padilla et al. 2007). It is now also clear that GPCRs can also signal from endosomes via Gα$_s$ proteins (Feinstein et al. 2013; Ferrandon et al. 2009; Irannejad et al. 2013; Thomsen et al. 2016; Tsvetanova et al. 2015; Van Dyke 2004). Endosomal Gα$_s$ signalling has also been reported for other GPCRs including the parathyroid hormone receptor (Ferrandon et al. 2009), β$_2$-adrenoceptor (Irannejad et al. 2013) and vasopressin type 2 receptor (Feinstein et al. 2013). Furthermore, super resolution microscopy revealed that β-arrestins and Gα$_s$ proteins remain physically associated with the GPCR in endosomes providing a platform to regulate cAMP formation (Feinstein et al. 2013). Additional interactions with the endosomal sorting proteins, GPCR-associated binding protein 1 and dysbindin, highlight a role for Gα$_s$ proteins in the post-endocytic sorting of receptors (Rosciglione et al. 2014).

A recent study has highlighted the importance of CGRP receptor-dependent endosomal signalling in generating distinct signals that are important for the transmission of pain (Yarwood et al. 2017). This study reaffirmed that the CGRP receptor is internalized in clathrin- and dynamin-dependent process and that it activates both PKA and PKC. In order to investigate the distinct signals generated from endosomes by the CGRP receptor, this study used a conjugation technique to target the CGRP receptor antagonist, $CGRP_{8\text{-}37}$, to endosomes (Jensen et al. 2017; Rajendran et al. 2008). $CGRP_{8\text{-}37}$ was conjugated to cholestenol using a polyethylene glycol 12 linker, promoting accumulation of the antagonist in endosomes and thus, providing a mechanism to antagonize CGRP receptor endosomal signalling. Incubation of CGRP receptor expressing cells with the cholesterol conjugated antagonist prevented activation of ERK in the nucleus, indicating a specific role for endosomal CGRP receptor signalling in generating ERK signals in the nucleus (Yarwood et al. 2017). The importance of this endosomal signalling for the transmission of pain was further investigated using capsaicin-, formalin- and complete Freund's adjuvant-induced model of mechanical allodynia. The study compared the effectiveness of $CGRP_{8\text{-}37}$ and endosomally targeted $CGRP_{8\text{-}37}$ in protecting against CGRP-induced pain and found that the unconjugated $CGRP_{8\text{-}37}$ afforded less protection than cholestenol-conjugated $CGRP_{8\text{-}37}$, supporting a role for CGRP receptor signalling from endosomes in the nociceptive process (Yarwood et al. 2017). It is interesting to note that this study also showed that the $G\alpha_s$ protein inhibitor, NF449 (Hulsmann et al. 2003), also suppressed nuclear ERK activation and thus, it is not clear if β-arrestin plays a role in this CGRP receptor endosomal signalling. It could be that the $G\alpha_s$ protein subunit remains associated with the active receptor in endosomes as has been shown for other receptors (Feinstein et al. 2013; Ferrandon et al. 2009; Irannejad et al. 2013) and that β-arrestin could mediate distinct and yet undiscovered endosomal signalling cascades.

6   Calcitonin Gene-Related Peptide-Mediated Signalling in Primary Cells

Although difficult to distinguish if all the effects of CGRP are mediated by CGRP receptors in primary cells and intact tissues due to the promiscuity of CGRP with other GPCRs, we will now examine the signalling cascades activated by CGRP in cells and tissues relevant in migraine. There are multiple sites where CGRP receptor may influence migraine: CGRP receptors in the cerebrovasculature (Edvinsson et al. 2002; Moreno et al. 1999; Oliver et al. 2002), CGRP receptors on dural mast cells (Ottosson and Edvinsson 1997; Theoharides et al. 2005), postsynaptic CGRP receptors on second-order sensory neurons (Fischer et al. 2005; Levy et al. 2005; Storer et al. 2004) and CGRP receptors in the trigeminal ganglion (Lennerz et al. 2008).

6.1   Calcitonin Gene-Related Peptide Signalling in Neuronal Cells

Exposure of cultured mouse trigeminal ganglion neurons cultures to CGRP promotes concentration-dependent increases in cAMP and also promotes up-regulation of CGRP itself (Zhang et al. 2007). Two PKA inhibitors, H89 and

8-Br-RP-cAMPS (Schafer et al. 1994), prevented CGRP-mediated effects on the CGRP promoter, providing consistent evidence with that observed in cell lines that CGRP couples to $G\alpha_s$ proteins, and promotes cAMP accumulation and activation of PKA. Supporting the activation of cAMP-dependent signalling, another study also in rat trigeminal ganglion neurons reported concentration-dependent increases in cAMP production in response to CGRP (Walker et al. 2015).

Further evidence for CGRP-dependent activation of protein kinases arises from a series of studies examining the regulation of $P2X_3$ receptors in mouse trigeminal ganglion cultures. $P2X_3$ receptors are activated by ATP and important in the transmission of pain signals [reviewed in Fabbretti (2013)]. Exposure of trigeminal ganglion cell cultures to GCRP promoted a delayed up-regulation of $P2X_3$ receptors at the cell-surface (Fabbretti et al. 2006). The cell-surface up-regulation of $P2X_3$ receptors was prevented by the CGRP receptor antagonist, $CGRP_{8-37}$, and by the protein kinase inhibitors, PKA inhibitor fragment 14–22 and chelerythrine chloride, implying a role for PKA and PKC, respectively (Fabbretti et al. 2006). In addition, CGRP treatment of mouse trigeminal ganglion cultures also enhances gene transcription of $P2X_3$ receptors (Simonetti et al. 2008), an observation supported by an independent study (Cady et al. 2011). In the former study, the authors observed CGRP-induced activation of Ca^{2+}-calmodulin-dependent kinase and phosphorylation of the cAMP-response element-binding protein, the latter presumably via a PKA-dependent mechanism. The effect of CGRP on gene transcription was only partially prevented by the inhibitors of PKA and PKC and co-incubation of the inhibitors did not have an additive effect, whereas KN93, an inhibitor of Ca^{2+}-calmodulin-dependent kinase, completely abolished the increase in transcription (Simonetti et al. 2008). Together, these studies provide evidence that CGRP may contribute to persistent pain through up-regulation of $P2X_3$ receptors and by redistribution of existing receptors to the cell surface. Brain-derived neurotrophic factor (BDNF) is known to play a role in processing pain through activity-dependent plastic changes in synaptic transmission [reviewed in Pezet and McMahon (2006)] and CGRP can promote the release of BDNF from trigeminal ganglion neurons (Buldyrev et al. 2006). A study showed that there are two distinct subpopulations of trigeminal ganglion neurons, expressing either BDNF receptors or CGRP receptors (32% overlap) (Simonetti et al. 2008). The distinct distribution of CGRP receptors and BNDF receptors in neurons raises the possibility that the BDNF released in response to CGRP can act in either an autocrine or paracrine manner. Similar to CGRP, BDNF promoted up-regulation of $P2X_3$ receptors is dependent on Ca^{2+}-calmodulin-dependent kinase (Simonetti et al. 2008).

Elevated actions of ERK and p38 are reported to be associated with neuronal sensitization and pain [reviewed in Ramesh (2014)]. To investigate the effect of CGRP on mitogen-activated protein kinases in neurons, rats were injected with CGRP in the temporomandibular joint and trigeminal ganglion isolated after 2 and 24 h (Cady et al. 2011). Examination of the isolated tissues by immunohistochemistry revealed that CGRP promoted activation of ERK, p38 and PKA both in neurons and satellite glial cells within the mandibular (V3) region of the ganglion, 24 h post-injection. Activation of sensory neurons has long been associated with expression of c-fos, a member of the intermediate family of transcription factors (Hunt et al. 1987). After injection of CGRP, the examination of rat spinal trigeminal nucleus revealed that CGRP promoted a significant up-regulation of c-fos expression both 2 and 24 h post-injection (Cady et al. 2011). There are multiple signalling mechanisms that promote the expression and activation of c-fos [reviewed in Gao and Ji (2009)]. However, ERK activity is known to promote the activation of cAMP-response element-binding protein, which in turn may bind to the promoter regions of many transcription factors including c-fos to induce their expression (Gille et al. 1995; Hodge et al. 1998; Sassone-Corsi et al. 1988). It is interesting to speculate that the CGRP-dependent expression of c-fos could be dependent on ERK activity, but this was not examined in this study.

Multiple CGRP-induced kinase signalling pathways have been postulated to provide neuroprotection of sensory, cortical and cerebellar neurons (Abushik et al. 2016). Cultured rat trigeminal, cortical and cerebellar neurons were exposed to homocysteine as a model of neurotoxicity [reviewed in Obeid and Herrmann (2006)]. Preincubation of neurons with CGRP for 20 min protected the neurons from the neurotoxic effect of homocysteine. The potential contributions of PKA (PKA inhibitor fragment 14–22), PKC (chelerythrine chloride) and Ca^{2+}-calmodulin-dependent kinase (KN93) pathways in this CGRP-mediated effect were examined using specific inhibitors of each pathway (Abushik et al. 2016). Incubation of neurons with KN93 and the PKA inhibitor, not the PKC inhibitor, prevented the protection afforded by CGRP. However, inhibition of PKC had no effect on CGRP-mediated neuronal protection. To provide proof that CGRP also protects neurons in vivo, mice were subjected to permanent middle cerebral artery occlusion as an ischaemic insult. Magnetic resonance images revealed significant reductions in the size of the lesion in mice treated with CGRP compared to control (Abushik et al. 2016). Unfortunately, the authors did not confirm the participation of the kinase pathways in the in vivo model.

6.2 Calcitonin Gene-Related Peptide Signalling in Glial Cells

Satellite glial cells surround the neuronal cell body and participate in signal processing and transmission in sensory ganglia [reviewed in Lecca et al. (2012)]. Nitric oxide (NO) promotes the relaxation of blood vessels, a phenomenon mimicked by CGRP and linked to the pathophysiology of migraine. CGRP receptors are also expressed on the satellite glial cells resident within the trigeminal ganglion (Eftekhari et al. 2010; Lennerz et al. 2008; Miller et al. 2016). Activation of glial cell cultures with CGRP promoted a concentration-dependent increase in the expression of inducible nitric oxide synthase (iNOS) and NO release (Li et al. 2008). A subsequent study by the same group showed that CGRP significantly increased the mitogen-activated protein kinases regulated transcription factors Elk, ATF-2 and CHOP (Vause and Durham 2009). In particular, the study highlighted increased CGRP-dependent activity of ERK, JNK and p38. Inhibition of each of the kinase pathway activities with specific kinase inhibitors [ERK, U0126 (inhibits an upstream kinase, mitogen-activated protein kinase kinase); JNK, SP600125 and p38, SB239063] prevented CGRP-induced up-regulation of iNOS and cellular production of NO (Vause and Durham 2009). Microarray analysis of CGRP-stimulated trigeminal glial cells also supports an important role for CGRP in regulating kinase-related signalling pathways (Vause and Durham 2010). In addition to kinase-related proteins, the microarray analysis also identified cytokines as a major target for CGRP-dependent induction. Stimulatory cytokines such as cytokine-induced neutrophil chemoattractant-3, fractalkine, granulocyte-macrophage colony-stimulating factor, interleukin (IL)-1α, leptin and macrophage inflammatory protein 3α having the largest up-regulation 8 h post-CGRP stimulation and induction of the inhibitory cytokines IL-10 and IL-4 was greatest after 24 h (Vause and Durham 2010).

Schwann cells support neuronal function by associating with nerve fibres and promote myelin production [reviewed in Miron (2017)]. The cell-surface components of the CGRP receptor, CLR and RAMP1, have been observed in Schwann cells in close proximity to both myelinated and unmyelinated nerves (Lennerz et al. 2008). A rat Schwann cell line (RT4-D6P2T) was stimulated with CGRP and a concentration-dependent induction of IL-1β was observed at protein level, by western blotting (Permpoonputtana et al. 2016). The production of IL-1β and IL-6 in response to CGRP increased over time, whereas there was no detectable increase in the production of tumour necrosis factor-α. The same study examined CGRP-induced kinase activation by western blotting, showing that CGRP promoted the phosphorylation of ERK. The phosphorylation of ERK was prevented by a CGRP receptor antagonist, $CGRP_{8-37}$, the PKA inhibitor, H89 and SQ22536, an inhibitor of adenylate cyclase (Permpoonputtana et al. 2016). Similarly, CGRP-induced expression of IL-1β was prevented by H89, SQ22536 and PD98059, an inhibitor of MEK-1. Together, this study showed the importance of the cAMP-PKA/ERK signalling cascade in the up-regulation of the important inflammatory mediator, IL-1β. Interestingly, unlike IL-1α which is expressed in cells under normal homeostasis, IL-1β is only expressed in response to inflammatory stimuli [reviewed in Dinarello (2011)], suggesting that CGRP is perceived as a danger signal by Schwann cells.

In an alternative approach to examining CGRP-induced signalling pathways in the trigeminal ganglion, the expression levels of inflammatory cytokines both at the mRNA level (using RT-PCR arrays and qPCR) and protein level (by immunohistochemistry) were analysed in an organ culture (Kristiansen and Edvinsson 2010). In this study, the authors observed that the morphological structure of the culture was well preserved after 24 h in culture and even in the absence of any stimulus there was a strong induction of pro-inflammatory cytokines including IL-6 and leukaemia inhibiting factor. Incubation of the organ culture with CGRP further enhanced the expression of a number of cytokines. Although this study did examine the effect of kinase inhibitors on expression of cytokines, they did not examine if they altered CGRP-dependent changes in cytokine expression (Kristiansen and Edvinsson 2010). Thus, it remains to be determined if CGRP activates kinase pathways to affect cytokine expression in an intact trigeminal ganglion culture.

Although it has been commonly reported that CGRP receptors promote mobilization of intracellular calcium, there is limited information as to whether CGRP promotes calcium mobilization in cells of the trigeminal ganglion. Cultures of mouse trigeminal ganglion were examined at the single cell level following exposure to CGRP using a fura2/AM-based assay (Ceruti et al. 2011). This study reported that although the satellite glial cells mobilized calcium in response to CGRP, this phenomenon was not observed in the trigeminal neurons and is supported by a previous study (Fabbretti et al. 2006). This evidence supports the notion that CGRP receptors are predominantly $G\alpha_s$-coupled in the neurons, but CGRP receptors expressed in the satellite glial cells are also coupled to $G\alpha_{q/11}$.

6.3 Calcitonin Gene-Related Peptide Signalling in Vascular Smooth Muscle Cells

CGRP receptors are present in the blood vessels that supply the brain. There have been many reports of expression of CGRP receptor components in vascular smooth muscle cells from different species (Cottrell et al. 2005; Lennerz et al. 2008; Miller et al. 2016). To date, there have not been any reports of expression of CGRP receptors on endothelial cells at sites relating to migraine. In agreement with this, CGRP-dependent relaxation of pial vessels from rabbit, cat and human was shown to be endothelium-independent (Hanko et al. 1985). The vascular smooth muscle cells are in close proximity to nerve endings, which contain CGRP. Therefore, nerves that release CGRP will activate smooth muscle CGRP receptors and thereby promote vasodilation via a neurogenic-dependent mechanism. In the periphery, CGRP and its receptors are present in endothelial cells (Doi et al. 2001; Hagner et al. 2001; Nikitenko et al. 2006) and an endothelium- and CGRP-dependent vasodilation has been reported in isolated rat aortic rings (Grace et al. 1987). In addition, CGRP promotes the proliferation of endothelial cells (Haegerstrand et al. 1990) and endothelial progenitor cells (Zhou et al. 2010). Although CGRP does not have a role on the regulation of blood pressure in normal individuals (Ho et al. 2010; Olesen et al. 2004), it does have a protective effect in individuals with cardiovascular disease [reviewed in Smillie and Brain (2011)]. In the absence of any studies on smooth muscle cells from migraine-related regions of the brain, we will discuss the studies investigating CGRP receptor signalling in vascular smooth muscle cells from other vascular beds.

The main blood supply to the cochlea is the spiral modiolar artery and altered blood flow in the cochlea was proposed to be involved in the development of hearing loss and tinnitus [reviewed in Hesse (2016)]. Thus, the activation of CGRP receptors in this artery was thought to be potentially beneficial. An investigation of the effect of CGRP on gerbil spiral modiolar artery responses measuring changes in vascular diameter also examined the potential signalling pathways involved (Herzog et al. 2002). Arteries were loaded with the calcium indicator dye, Fluo-4, and changes in the calcium levels in response to a challenge with CGRP were monitored by microscopy. CGRP was found to cause a transient decrease in intracellular calcium levels of the smooth muscle cells. Furthermore,

CGRP also stimulated cAMP production (Herzog et al. 2002), suggesting that CGRP receptors in smooth muscle cells are coupled to $G\alpha_s$ proteins. The decrease in calcium concentration was speculated to be attributable to the opening of K^+ channels causing hyperpolarization and closure of L-type Ca^{2+} channels and a cAMP-dependent effect on the contractile apparatus (Herzog et al. 2002).

Scherer et al. reported the reversal of endothelin-1 induced spasms of spiral modiolar artery by CGRP (Scherer et al. 2002). The molecular basis for this CGRP-specific reversal has since been investigated in mesenteric arteries (Meens et al. 2009, 2010, 2011, 2012), that also express CGRP receptors with activation causing vessel relaxation (Cottrell et al. 2005; Lei et al. 1994). Endothelin-1 is potent vasoconstrictor (O'Brien et al. 1987; Yanagisawa et al. 1988) and can activate both endothelin A (ET_A) receptors and endothelin B receptors (Arai et al. 1990; Sakurai et al. 1990) present on vascular smooth muscle cells (Hori et al. 1992; Nakamichi et al. 1992; Russell et al. 1997; Wendel-Wellner et al. 2002). In the rat mesenteric arteries, endothelin-1 promoted a potent concentration-dependent contraction that was prevented by the ET_A receptor antagonist, BQ123, but not by the endothelin B receptor antagonist, BQ788, indicating that only ET_A receptors were involved in the endothelin-1-dependent effect (Meens et al. 2009). Capsaicin acts on transient potential vanilloid 1 receptors to release CGRP from nerve endings, promoting the relaxation of mesenteric arteries (Fujimori et al. 1990). Stimulation of mesenteric arteries with capsaicin promoted a significantly greater relaxation of mesenteric arteries that were pre-contracted with endothelin-1, compared to arteries pre-contracted with phenylephrine with the authors speculating that endothelin-1-promoted contractions are hypersensitive to CGRP-mediated reversal (Meens et al. 2009). In contrast, other relaxing agents including isoproterenol (a β_2-adrenoceptor agonist and activator of cAMP production), forskolin (a direct activator of adenylate cyclase), sodium nitroprusside (an NO donor) and pinacidil (an activator of ATP-sensitive K^+-channels) all caused similar concentration-dependent relaxations irrespective of whether endothelin-1 or phenylephrine were used to cause the initial contraction (Meens et al. 2009). Investigation into the mechanism by which activation of CGRP receptors selectively reverses the effects of endothelin-1 on ET_A receptors identified a previously unidentified CGRP receptor-dependent signalling pathway (Meens et al. 2012). The long-lasting effect of endothelin-1 on the contractile responses of mesenteric arteries is attributed to the tight binding and slow dissociation kinetics of endothelin-1 on ET_A receptors [reviewed in De Mey et al. (2009, 2011)]. Using fluorescently labelled endothelin-1 as an agonist of ET_A receptor and two-photon laser scanning microscopy, CGRP was observed to promote the dissociation of endothelin-1 from the smooth muscle layer of mesenteric arteries (Meens et al. 2010). As other activators of cAMP production did not mimic this ligand/receptor dissociation, the authors investigated the role of $G\beta\gamma$ protein subunits (Meens et al. 2012). The study showed that the CGRP-specific effect on endothelin-1 contractions was independent of cAMP, as inhibitors of adenylate cyclase (SQ22536) and PKA (H89 and KT5720) were ineffective at blocking the CGRP-dependent reversal of endothelin-promoted contractions (Meens et al. 2012). However, gallein, a low molecular mass inhibitor of $G\beta\gamma$ protein function (Lehmann et al. 2008), not only enhanced CGRP-induced cAMP production in cultured rat mesenteric vascular smooth muscle cells, it also prevented the CGRP-specific effect of relaxing endothelin-1-dependent contractions (Meens et al. 2012). To date, this remains the only report of a function of CGRP-activated $G\beta\gamma$ protein subunits, although if any $G\beta\gamma$ proteins are involved remains to be determined.

Migraines have been proposed as a reaction to the higher levels of oxidative stress that occur between attacks in migraineurs [reviewed in Borkum (2018)]. This raises the question of whether CGRP is released as a consequence of and a reaction to oxidative stress. Sensory nerves express a plethora of transient receptor potential ion channels including transient receptor potential ankyrin 1 receptor (TRPA1) (Kobayashi et al. 2005; Story et al. 2003). TRPA1 is activated by a variety of endogenous and exogenous agents including mustard oil (McNamara et al. 2007), 4-hydroxynonenal (Macpherson et al. 2007; Trevisani et al. 2007) and allicin (Bautista et al. 2005; Macpherson et al. 2005). In addition, agents that generate oxidative stress also promote the activation of TRPA1 (Hill and Schaefer 2009). Thus, as activation of TRPA1 channels promotes the release of CGRP (Trevisani et al. 2007), the release of CGRP in migraineurs may be a consequence of the increased oxidative stress. However, a protective role for CGRP against oxidative stress has been proposed (Schaeffer et al. 2003a, b). Vascular smooth muscle cells were exposed to oxidative stress using a hydrogen peroxide generating system (glucose/glucose oxidase system) for 1 h and the viability of cells then examined using an MTT assay and apoptosis examined using Hoechst staining and annexin-V labelling (Schaeffer et al. 2003b). The viability was diminished in a concentration-dependent manner and cells were rescued by pretreatment of cell with CGRP. Investigation of the signalling pathways involved observed that although the generation of oxidative stress itself increased the activity of ERK, pretreatment of the cells with CGRP significantly enhanced the ERK activity (Schaeffer et al. 2003b). No role for JNK in CGRP-dependent prevention of the hydrogen peroxide-induced reduction in cell viability was observed (Schaeffer et al. 2003b). In contrast, inhibitors of ERK (PD98059) and p38 (SB203580) both reversed the effect of CGRP protection (Schaeffer et al. 2003b). In support of the protective effect of CGRP against oxidative stress, lentivirus-mediated overexpression of CGRP in the Schwann cell line, RSC96, transiently protected against high levels of glucose in the growth medium (Wu et al. 2015). The CGRP-induced signalling pathway in this cell line remains unidentified. Therefore, during high levels of oxidative stress CGRP may be released to maintain cell viability, but is it other signalling events that contribute the sensitization of sensory neurons and increased vasodilation, that are deleterious and contribute to the development of migraine.

7 Conclusions

It is clear that we have discovered a wealth of information regarding the signalling mediated by both CGRP and CGRP receptors and that it is complex, yet compared with other GPCRs much less of known about these signalling pathways. We have delineated some of the CGRP and CGRP receptor pathways and it is now known that they involve multiple types of G protein, protein kinases and transcription factors. Understanding the role of CGRP is complicated in that some of the effects of CGRP have been shown to be protective, whereas others have deleterious effects and contribute to the development of migraine pathophysiology. Preventing the activation of CGRP receptor (with small molecule inhibitors and monoclonal antibodies) and the actions of CGRP (with monoclonal antibodies) have proved that CGRP and its receptor are valid targets for the prevention of migraine. The relatively new avenue of CGRP receptor signalling along the endocytic pathway and the involvement of these distinct signalling pathways in the transmission of pain is an exciting development. This phenomenon requires further investigation and will no doubt prove vital for identifying new potential targets. Whether these targets will be known or novel proteins, or whether they are allosteric modulators or biased agonists or antagonists of CGRP receptors is an unknown commodity. Only further knowledge and scientific understanding of CGRP- and CGRP receptor-mediated signalling pathways will help in the quest to make migraine pain a problem of the past.

References

Abushik PA et al (2016) Pro-nociceptive migraine mediator CGRP provides neuroprotection of sensory, cortical and cerebellar neurons via multi-kinase signaling. Cephalalgia. https://doi.org/10.1177/0333102416681588

Aiyar N, Disa J, Dang K, Pronin AN, Benovic JL, Nambi P (2000) Involvement of G protein-coupled receptor kinase-6 in desensitization of CGRP receptors. Eur J Pharmacol 403:1–7

Aiyar N, Disa J, Stadel JM, Lysko PG (1999) Calcitonin gene-related peptide receptor independently stimulates $3',5'$-cyclic adenosine monophosphate and Ca^{2+} signaling pathways. Mol Cell Biochem 197:179–185

Amara SG, Jonas V, Rosenfeld MG, Ong ES, Evans RM (1982) Alternative RNA processing in calcitonin gene expression generates mRNAs encoding different polypeptide products. Nature 298:240–244

Arai H, Hori S, Aramori I, Ohkubo H, Nakanishi S (1990) Cloning and expression of a cDNA encoding an endothelin receptor. Nature 348:730–732. https://doi.org/10.1038/348730a0

Bache KG, Raiborg C, Mehlum A, Stenmark H (2003) STAM and Hrs are subunits of a multivalent ubiquitin-binding complex on early endosomes. J Biol Chem 278:12513–12521. https://doi.org/10.1074/jbc.M210843200

Banerjee S, Evanson J, Harris E, Lowe SL, Thomasson KA, Porter JE (2006) Identification of specific calcitonin-like receptor residues important for calcitonin gene-related peptide high affinity binding. BMC Pharmacol 6:9. https://doi.org/10.1186/1471-2210-6-9

Bautista DM et al (2005) Pungent products from garlic activate the sensory ion channel TRPA1. Proc Natl Acad Sci U S A 102:12248–12252. https://doi.org/10.1073/pnas.0505356102

Borkum JM (2018) The migraine attack as a homeostatic, neuroprotective response to brain oxidative stress: preliminary evidence for a theory. Headache 58:118–135. https://doi.org/10.1111/head.13214

Budenholzer L, Cheng CL, Li Y, Hochstrasser M (2017) Proteasome structure and assembly. J Mol Biol 429:3500–3524. https://doi.org/10.1016/j.jmb.2017.05.027

Buldyrev I, Tanner NM, Hsieh HY, Dodd EG, Nguyen LT, Balkowiec A (2006) Calcitonin gene-related peptide enhances release of native brain-derived neurotrophic factor from trigeminal ganglion neurons. J Neurochem 99:1338–1350. https://doi.org/10.1111/j.1471-4159.2006.04161.x

Cady RJ, Glenn JR, Smith KM, Durham PL (2011) Calcitonin gene-related peptide promotes cellular changes in trigeminal neurons and glia implicated in peripheral and central sensitization. Mol Pain 7:94. https://doi.org/10.1186/1744-8069-7-94

Ceruti S et al (2011) Calcitonin gene-related peptide-mediated enhancement of purinergic neuron/glia communication by the algogenic factor bradykinin in mouse trigeminal ganglia from wild-type and R192Q Cav2.1 knock-in mice: implications for basic mechanisms of migraine pain. J Neurosci 31:3638–3649. https://doi.org/10.1523/JNEUROSCI.6440-10.2011

Cottrell GS, Padilla B, Pikios S, Roosterman D, Steinhoff M, Grady EF, Bunnett NW (2007) Post-endocytic sorting of calcitonin receptor-like receptor and receptor activity-modifying protein 1. J Biol Chem 282:12260–12271. https://doi.org/10.1074/jbc.M606338200

Cottrell GS et al (2009) Endosomal endothelin-converting enzyme-1: a regulator of beta-arrestin-dependent ERK signaling. J Biol Chem 284:22411–22425. https://doi.org/10.1074/jbc.M109.026674

Cottrell GS et al (2005) Localization of calcitonin receptor-like receptor and receptor activity modifying protein 1 in enteric neurons, dorsal root ganglia, and the spinal cord of the rat. J Comp Neurol 490:239–255. https://doi.org/10.1002/cne.20669

De Mey JG, Compeer MG, Lemkens P, Meens MJ (2011) ETA-receptor antagonists or allosteric modulators? Trends Pharmacol Sci 32:345–351. https://doi.org/10.1016/j.tips.2011.02.018

De Mey JG, Compeer MG, Meens MJ (2009) Endothelin-1, an endogenous irreversible agonist in search of an allosteric inhibitor. Mol Cell Pharmacol 1:246–257

DeFea KA, Vaughn ZD, O'Bryan EM, Nishijima D, Dery O, Bunnett NW (2000) The proliferative and antiapoptotic effects of substance P are facilitated by formation of a beta-arrestin-dependent scaffolding complex. Proc Natl Acad Sci U S A 97:11086–11091. https://doi.org/10.1073/pnas.190276697

Dinarello CA (2011) Interleukin-1 in the pathogenesis and treatment of inflammatory diseases. Blood 117:3720–3732. https://doi.org/10.1182/blood-2010-07-273417

Doi Y et al (2001) Synthesis of calcitonin gene-related peptide (CGRP) by rat arterial endothelial cells. Histol Histopathol 16:1073–1079

Edvinsson L, Alm R, Shaw D, Rutledge RZ, Koblan KS, Longmore J, Kane SA (2002) Effect of the CGRP receptor antagonist BIBN4096BS in human cerebral, coronary and omental arteries and in SK-N-MC cells. Eur J Pharmacol 434:49–53

Edvinsson L, Fredholm BB, Hamel E, Jansen I, Verrecchia C (1985) Perivascular peptides relax cerebral arteries concomitant with stimulation of cyclic adenosine monophosphate accumulation or release of an endothelium-derived relaxing factor in the cat. Neurosci Lett 58:213–217

Eftekhari S, Salvatore CA, Calamari A, Kane SA, Tajti J, Edvinsson L (2010) Differential distribution of calcitonin gene-related peptide and its receptor components in the human trigeminal ganglion. Neuroscience 169:683–696. https://doi.org/10.1016/j.neuroscience.2010.05.016

Egea SC, Dickerson IM (2012) Direct interactions between calcitonin-like receptor (CLR) and CGRP-receptor component protein (RCP) regulate CGRP receptor signaling. Endocrinology 153:1850–1860. https://doi.org/10.1210/en.2011-1459

Eichel K, von Zastrow M (2018) Subcellular organization of GPCR signaling. Trends Pharmacol Sci 39:200–208. https://doi.org/10.1016/j.tips.2017.11.009

Evans BN, Rosenblatt MI, Mnayer LO, Oliver KR, Dickerson IM (2000) CGRP-RCP, a novel protein required for signal transduction at calcitonin gene-related peptide and adrenomedullin receptors. J Biol Chem 275:31438–31443. https://doi.org/10.1074/jbc.M005604200

Fabbretti E (2013) ATP P2X3 receptors and neuronal sensitization. Front Cell Neurosci 7:236. https://doi.org/10.3389/fncel.2013.00236

Fabbretti E, D'Arco M, Fabbro A, Simonetti M, Nistri A, Giniatullin R (2006) Delayed upregulation of ATP P2X3 receptors of trigeminal sensory neurons by calcitonin gene-related peptide. J Neurosci 26:6163–6171. https://doi.org/10.1523/JNEUROSCI.0647-06.2006

Fahnoe DC, Knapp J, Johnson GD, Ahn K (2000) Inhibitor potencies and substrate preference for endothelin-converting enzyme-1 are dramatically affected by pH. J Cardiovasc Pharmacol 36:S22–S25

Feinstein TN et al (2013) Noncanonical control of vasopressin receptor type 2 signaling by retromer and arrestin. J Biol Chem 288:27849–27860. https://doi.org/10.1074/jbc.M112.445098

Ferrandon S et al (2009) Sustained cyclic AMP production by parathyroid hormone receptor endocytosis. Nat Chem Biol 5:734–742. https://doi.org/10.1038/nchembio.206

Fischer MJ, Koulchitsky S, Messlinger K (2005) The nonpeptide calcitonin gene-related peptide receptor antagonist BIBN4096BS lowers the activity of neurons with meningeal input in the rat spinal trigeminal nucleus. J Neurosci 25:5877–5883. https://doi.org/10.1523/JNEUROSCI.0869-05.2005

Fraser NJ, Wise A, Brown J, McLatchie LM, Main MJ, Foord SM (1999) The amino terminus of receptor activity modifying proteins is a critical determinant of glycosylation state and ligand binding of calcitonin receptor-like receptor. Mol Pharmacol 55:1054–1059

Fujimori A, Saito A, Kimura S, Goto K (1990) Release of calcitonin gene-related peptide (CGRP) from capsaicin-sensitive vasodilator nerves in the rat mesenteric artery. Neurosci Lett 112:173–178

Gao YJ, Ji RR (2009) c-Fos and pERK, which is a better marker for neuronal activation and central sensitization after noxious stimulation and tissue injury? Open Pain J 2:11–17. https://doi.org/10.2174/1876386300902010011

Gille H, Kortenjann M, Thomae O, Moomaw C, Slaughter C, Cobb MH, Shaw PE (1995) ERK phosphorylation potentiates Elk-1-mediated ternary complex formation and transactivation. EMBO J 14:951–962

Grace GC, Dusting GJ, Kemp BE, Martin TJ (1987) Endothelium and the vasodilator action of rat calcitonin gene-related peptide (CGRP). Br J Pharmacol 91:729–733

Guillemare E, Lazdunski M, Honore E (1994) CGRP-induced activation of KATP channels in follicular Xenopus oocytes. Pflugers Arch 428:604–609

Gurevich VV, Gurevich EV (2015) Analyzing the roles of multi-functional proteins in cells: the case of arrestins and GRKs. Crit Rev Biochem Mol Biol 50:440–452. https://doi.org/10.3109/10409238.2015.1067185

Haegerstrand A, Dalsgaard CJ, Jonzon B, Larsson O, Nilsson J (1990) Calcitonin gene-related peptide stimulates proliferation of human endothelial cells. Proc Natl Acad Sci U S A 87:3299–3303

Hagner S et al (2001) Immunohistochemical detection of calcitonin gene-related peptide receptor (CGRPR)-1 in the endothelium of human coronary artery and bronchial blood vessels. Neuropeptides 35:58–64. https://doi.org/10.1054/npep.2000.0844

Hanko J, Hardebo JE, Kahrstrom J, Owman C, Sundler F (1985) Calcitonin gene-related peptide is present in mammalian cerebrovascular nerve fibres and dilates pial and peripheral arteries. Neurosci Lett 57:91–95

Hanyaloglu AC, McCullagh E, von Zastrow M (2005) Essential role of Hrs in a recycling mechanism mediating functional resensitization of cell signaling. EMBO J 24:2265–2283. https://doi.org/10.1038/sj.emboj.7600688

Hasdemir B, Bunnett NW, Cottrell GS (2007) Hepatocyte growth factor-regulated tyrosine kinase substrate (HRS) mediates post-endocytic trafficking of protease-activated receptor 2 and calcitonin receptor-like receptor. J Biol Chem 282:29646–29657. https://doi.org/10.1074/jbc.M702974200

Hay DL, Pioszak AA (2016) Receptor activity-modifying proteins (RAMPs): new insights and roles. Annu Rev Pharmacol Toxicol 56:469–487. https://doi.org/10.1146/annurev-pharmtox-010715-103120

Hershko A, Ciechanover A, Heller H, Haas AL, Rose IA (1980) Proposed role of ATP in protein breakdown: conjugation of protein with multiple chains of the polypeptide of ATP-dependent proteolysis. Proc Natl Acad Sci U S A 77:1783–1786

Herskovits JS, Burgess CC, Obar RA, Vallee RB (1993) Effects of mutant rat dynamin on endocytosis. J Cell Biol 122:565–578

Herzog M, Scherer EQ, Albrecht B, Rorabaugh B, Scofield MA, Wangemann P (2002) CGRP receptors in the gerbil spiral modiolar artery mediate a sustained vasodilation via a transient cAMP-mediated Ca^{2+}-decrease. J Membr Biol 189:225–236. https://doi.org/10.1007/s00232-002-1017-5

Hesse G (2016) Evidence and evidence gaps in tinnitus therapy. GMS Curr Top Otorhinolaryngol Head Neck Surg 15:Doc04. https://doi.org/10.3205/cto000131

Hilairet S, Belanger C, Bertrand J, Laperriere A, Foord SM, Bouvier M (2001) Agonist-promoted internalization of a ternary complex between calcitonin receptor-like receptor, receptor activity-modifying protein 1 (RAMP1), and beta-arrestin. J Biol Chem 276:42182–42190. https://doi.org/10.1074/jbc.M107323200

Hill K, Schaefer M (2009) Ultraviolet light and photosensitising agents activate TRPA1 via generation of oxidative stress. Cell Calcium 45:155–164. https://doi.org/10.1016/j.ceca.2008.08.001

Hislop JN, Marley A, Von Zastrow M (2004) Role of mammalian vacuolar protein-sorting proteins in endocytic trafficking of a non-ubiquitinated G protein-coupled receptor to lysosomes. J Biol Chem 279:22522–22531. https://doi.org/10.1074/jbc.M311062200

Ho AP et al (2010) Randomized, controlled trial of telcagepant over four migraine attacks. Cephalalgia 30:1443–1457. https://doi.org/10.1177/0333102410370878

Hodge C, Liao J, Stofega M, Guan K, Carter-Su C, Schwartz J (1998) Growth hormone stimulates phosphorylation and activation of elk-1 and expression of c-fos, egr-1, and junB through activation of extracellular signal-regulated kinases 1 and 2. J Biol Chem 273:31327–31336

Hori S, Komatsu Y, Shigemoto R, Mizuno N, Nakanishi S (1992) Distinct tissue distribution and cellular localization of two messenger ribonucleic acids encoding different subtypes of rat endothelin receptors. Endocrinology 130:1885–1895. https://doi.org/10.1210/endo.130.4.1312429

Hulsmann M, Nickel P, Kassack M, Schmalzing G, Lambrecht G, Markwardt F (2003) NF449, a novel picomolar potency antagonist at human P2X1 receptors. Eur J Pharmacol 470:1–7

Hunt SP, Pini A, Evan G (1987) Induction of c-fos-like protein in spinal cord neurons following sensory stimulation. Nature 328:632–634. https://doi.org/10.1038/328632a0

Irannejad R et al (2013) Conformational biosensors reveal GPCR signalling from endosomes. Nature 495:534–538. https://doi.org/10.1038/nature12000

Irannejad R, von Zastrow M (2014) GPCR signaling along the endocytic pathway. Curr Opin Cell Biol 27:109–116. https://doi.org/10.1016/j.ceb.2013.10.003

Jacob C, Cottrell GS, Gehringer D, Schmidlin F, Grady EF, Bunnett NW (2005) c-Cbl mediates ubiquitination, degradation, and down-regulation of human protease-activated receptor 2. J Biol Chem 280:16076–16087. https://doi.org/10.1074/jbc.M500109200

Jean-Charles PY, Kaur S, Shenoy SK (2017) G protein-coupled receptor signaling through beta-arrestin-dependent mechanisms. J Cardiovasc Pharmacol 70:142–158. https://doi.org/10.1097/FJC.0000000000000482

Jensen DD et al (2017) Neurokinin 1 receptor signaling in endosomes mediates sustained nociception and is a viable therapeutic target for prolonged pain relief. Sci Transl Med 9. https://doi.org/10.1126/scitranslmed.aal3447

Klein KR, Matson BC, Caron KM (2016) The expanding repertoire of receptor activity modifying protein (RAMP) function. Crit Rev Biochem Mol Biol 51:65–71. https://doi.org/10.3109/10409238.2015.1128875

Klein SL, Strausberg RL, Wagner L, Pontius J, Clifton SW, Richardson P (2002) Genetic and genomic tools for Xenopus research: the NIH Xenopus initiative. Dev Dyn 225:384–391. https://doi.org/10.1002/dvdy.10174

Kline LW, Kaneko T, Chiu KW, Harvey S, Pang PK (1988) Calcitonin gene-related peptide in the bullfrog, Rana catesbeiana: localization and vascular actions. Gen Comp Endocrinol 72:123–129

Kobayashi H et al (1987) Calcitonin gene related peptide stimulates adenylate cyclase activity in rat striated muscle. Experientia 43:314–316

Kobayashi K, Fukuoka T, Obata K, Yamanaka H, Dai Y, Tokunaga A, Noguchi K (2005) Distinct expression of TRPM8, TRPA1, and TRPV1 mRNAs in rat primary afferent neurons with adelta/c-fibers and colocalization with trk receptors. J Comp Neurol 493:596–606. https://doi.org/10.1002/cne.20794

Komada M, Masaki R, Yamamoto A, Kitamura N (1997) Hrs, a tyrosine kinase substrate with a conserved double zinc finger domain, is localized to the cytoplasmic surface of early endosomes. J Biol Chem 272:20538–20544

Kristiansen KA, Edvinsson L (2010) Neurogenic inflammation: a study of rat trigeminal ganglion. J Headache Pain 11:485–495. https://doi.org/10.1007/s10194-010-0260-x

Kuwasako K, Kitamura K, Nagoshi Y, Cao YN, Eto T (2003) Identification of the human receptor activity-modifying protein 1 domains responsible for agonist binding specificity. J Biol Chem 278:22623–22630. https://doi.org/10.1074/jbc.M302571200

Kuwasako K et al (2000) Visualization of the calcitonin receptor-like receptor and its receptor activity-modifying proteins during internalization and recycling. J Biol Chem 275:29602–29609. https://doi.org/10.1074/jbc.M004534200

Laufer R, Changeux JP (1987) Calcitonin gene-related peptide elevates cyclic AMP levels in chick skeletal muscle: possible neurotrophic role for a coexisting neuronal messenger. EMBO J 6:901–906

Lecca D, Ceruti S, Fumagalli M, Abbracchio MP (2012) Purinergic trophic signalling in glial cells: functional effects and modulation of cell proliferation, differentiation, and death. Purinergic Signal 8:539–557. https://doi.org/10.1007/s11302-012-9310-y

Lehmann DM, Seneviratne AM, Smrcka AV (2008) Small molecule disruption of G protein beta gamma subunit signaling inhibits neutrophil chemotaxis and inflammation. Mol Pharmacol 73:410–418. https://doi.org/10.1124/mol.107.041780

Lei S, Mulvany MJ, Nyborg NC (1994) Characterization of the CGRP receptor and mechanisms of action in rat mesenteric small arteries. Pharmacol Toxicol 74:130–135

Lennerz JK, Ruhle V, Ceppa EP, Neuhuber WL, Bunnett NW, Grady EF, Messlinger K (2008) Calcitonin receptor-like receptor (CLR), receptor activity-modifying protein 1 (RAMP1), and calcitonin gene-related peptide (CGRP) immunoreactivity in the rat trigeminovascular system: differences between peripheral and central CGRP receptor distribution. J Comp Neurol 507:1277–1299. https://doi.org/10.1002/cne.21607

Levy D, Burstein R, Strassman AM (2005) Calcitonin gene-related peptide does not excite or sensitize meningeal nociceptors: implications for the pathophysiology of migraine. Ann Neurol 58:698–705. https://doi.org/10.1002/ana.20619

Li J, Vause CV, Durham PL (2008) Calcitonin gene-related peptide stimulation of nitric oxide synthesis and release from trigeminal ganglion glial cells. Brain Res 1196:22–32. https://doi.org/10.1016/j.brainres.2007.12.028

Luebke AE, Dahl GP, Roos BA, Dickerson IM (1996) Identification of a protein that confers calcitonin gene-related peptide responsiveness to oocytes by using a cystic fibrosis transmembrane conductance regulator assay. Proc Natl Acad Sci U S A 93:3455–3460

Luttrell LM et al (1999) Beta-arrestin-dependent formation of beta2 adrenergic receptor-Src protein kinase complexes. Science 283:655–661

Macpherson LJ, Dubin AE, Evans MJ, Marr F, Schultz PG, Cravatt BF, Patapoutian A (2007) Noxious compounds activate TRPA1 ion channels through covalent modification of cysteines. Nature 445:541–545. https://doi.org/10.1038/nature05544

Macpherson LJ, Geierstanger BH, Viswanath V, Bandell M, Eid SR, Hwang S, Patapoutian A (2005) The pungency of garlic: activation of TRPA1 and TRPV1 in response to allicin. Curr Biol 15:929–934. https://doi.org/10.1016/j.cub.2005.04.018

Main MJ, Brown J, Brown S, Fraser NJ, Foord SM (1998) The CGRP receptor can couple via pertussis toxin sensitive and insensitive G proteins. FEBS Lett 441:6–10

Marchese A, Benovic JL (2001) Agonist-promoted ubiquitination of the G protein-coupled receptor CXCR4 mediates lysosomal sorting. J Biol Chem 276:45509–45512. https://doi.org/10.1074/jbc.C100527200

McLatchie LM et al (1998) RAMPs regulate the transport and ligand specificity of the calcitonin-receptor-like receptor. Nature 393:333–339. https://doi.org/10.1038/30666

McNamara CR et al (2007) TRPA1 mediates formalin-induced pain. Proc Natl Acad Sci U S A 104:13525–13530. https://doi.org/10.1073/pnas.0705924104

McNeish AJ, Roux BT, Aylett SB, Van Den Brink AM, Cottrell GS (2012) Endosomal proteolysis regulates calcitonin gene-related peptide responses in mesenteric arteries. Br J Pharmacol 167:1679–1690. https://doi.org/10.1111/j.1476-5381.2012.02129.x

Meens MJ, Compeer MG, Hackeng TM, van Zandvoort MA, Janssen BJ, De Mey JG (2010) Stimuli of sensory-motor nerves terminate arterial contractile effects of endothelin-1 by CGRP and dissociation of ET-1/ET(A)-receptor complexes. PLoS One 5:e10917. https://doi.org/10.1371/journal.pone.0010917

Meens MJ, Fazzi GE, van Zandvoort MA, De Mey JG (2009) Calcitonin gene-related peptide selectively relaxes contractile responses to endothelin-1 in rat mesenteric resistance arteries. J Pharmacol Exp Ther 331:87–95. https://doi.org/10.1124/jpet.109.155143

Meens MJ, Mattheij NJ, Nelissen J, Lemkens P, Compeer MG, Janssen BJ, De Mey JG (2011) Calcitonin gene-related peptide terminates long-lasting vasopressor responses to endothelin 1 in vivo. Hypertension 58:99–106. https://doi.org/10.1161/HYPERTENSIONAHA.110.169128

Meens MJ et al (2012) G-protein betagamma subunits in vasorelaxing and anti-endothelinergic effects of calcitonin gene-related peptide. Br J Pharmacol 166:297–308. https://doi.org/10.1111/j.1476-5381.2011.01774.x

Miller S, Liu H, Warfvinge K, Shi L, Dovlatyan M, Xu C, Edvinsson L (2016) Immunohistochemical localization of the calcitonin gene-related peptide binding site in the primate trigeminovascular system using functional antagonist antibodies. Neuroscience 328:165–183. https://doi.org/10.1016/j.neuroscience.2016.04.046

Miron VE (2017) Microglia-driven regulation of oligodendrocyte lineage cells, myelination, and remyelination. J Leukoc Biol 101:1103–1108. https://doi.org/10.1189/jlb.3RI1116-494R

Miyauchi K, Tadotsu N, Hayashi T, Ono Y, Tokoyoda K, Tsujikawa K, Yamamoto H (2002) Molecular cloning and characterization of mouse calcitonin gene-related peptide receptor. Neuropeptides 36:22–33

Moreno MJ, Cohen Z, Stanimirovic DB, Hamel E (1999) Functional calcitonin gene-related peptide type 1 and adrenomedullin receptors in human trigeminal ganglia, brain vessels, and cerebromicrovascular or astroglial cells in culture. J Cereb Blood Flow Metab 19:1270–1278. https://doi.org/10.1097/00004647-199911000-00012

Nakamichi K, Ihara M, Kobayashi M, Saeki T, Ishikawa K, Yano M (1992) Different distribution of endothelin receptor subtypes in pulmonary tissues revealed by the novel selective ligands BQ-123 and [Ala1,3,11,15]ET-1. Biochem Biophys Res Commun 182:144–150

Nikitenko LL, Blucher N, Fox SB, Bicknell R, Smith DM, Rees MC (2006) Adrenomedullin and CGRP interact with endogenous calcitonin-receptor-like receptor in endothelial cells and induce its desensitisation by different mechanisms. J Cell Sci 119:910–922. https://doi.org/10.1242/jcs.02783

O'Brien RF, Robbins RJ, McMurtry IF (1987) Endothelial cells in culture produce a vasoconstrictor substance. J Cell Physiol 132:263–270. https://doi.org/10.1002/jcp.1041320210

Oakley RH, Laporte SA, Holt JA, Caron MG, Barak LS (2000) Differential affinities of visual arrestin, beta arrestin1, and beta arrestin2 for G protein-coupled receptors delineate two major classes of receptors. J Biol Chem 275:17201–17210. https://doi.org/10.1074/jbc.M910348199

Obeid R, Herrmann W (2006) Mechanisms of homocysteine neurotoxicity in neurodegenerative diseases with special reference to dementia. FEBS Lett 580:2994–3005. https://doi.org/10.1016/j.febslet.2006.04.088

Olesen J et al (2004) Calcitonin gene-related peptide receptor antagonist BIBN 4096 BS for the acute treatment of migraine. N Engl J Med 350:1104–1110. https://doi.org/10.1056/NEJMoa030505

Oliver KR, Wainwright A, Edvinsson L, Pickard JD, Hill RG (2002) Immunohistochemical localization of calcitonin receptor-like receptor and receptor activity-modifying proteins in the human cerebral vasculature. J Cereb Blood Flow Metab 22:620–629. https://doi.org/10.1097/00004647-200205000-00014

Ottosson A, Edvinsson L (1997) Release of histamine from dural mast cells by substance P and calcitonin gene-related peptide. Cephalalgia 17:166–174. https://doi.org/10.1046/j.1468-2982.1997.1703166.x

Padilla BE, Cottrell GS, Roosterman D, Pikios S, Muller L, Steinhoff M, Bunnett NW (2007) Endothelin-converting enzyme-1 regulates endosomal sorting of calcitonin receptor-like receptor and beta-arrestins. J Cell Biol 179:981–997. https://doi.org/10.1083/jcb.200704053

Parameswaran N, Disa J, Spielman WS, Brooks DP, Nambi P, Aiyar N (2000) Activation of multiple mitogen-activated protein kinases by recombinant calcitonin gene-related peptide receptor. Eur J Pharmacol 389:125–130

Permpoonputtana K, Porter JE, Govitrapong P (2016) Calcitonin gene-related peptide mediates an inflammatory response in Schwann cells via cAMP-dependent ERK signaling cascade. Life Sci 144:19–25. https://doi.org/10.1016/j.lfs.2015.11.015

Peterson YK, Luttrell LM (2017) The diverse roles of arrestin scaffolds in G protein-coupled receptor signaling. Pharmacol Rev 69:256–297. https://doi.org/10.1124/pr.116.013367

Pezet S, McMahon SB (2006) Neurotrophins: mediators and modulators of pain. Annu Rev Neurosci 29:507–538. https://doi.org/10.1146/annurev.neuro.29.051605.112929

Prado MA, Evans-Bain B, Oliver KR, Dickerson IM (2001) The role of the CGRP-receptor component protein (RCP) in adrenomedullin receptor signal transduction. Peptides 22:1773–1781

Rajendran L et al (2008) Efficient inhibition of the Alzheimer's disease beta-secretase by membrane targeting. Science 320:520–523. https://doi.org/10.1126/science.1156609

Ramesh G (2014) Novel therapeutic targets in neuroinflammation and neuropathic pain. Inflamm Cell Signal 1. https://doi.org/10.14800/ics.111

Rosciglione S, Theriault C, Boily MO, Paquette M, Lavoie C (2014) Galphas regulates the post-endocytic sorting of G protein-coupled receptors. Nat Commun 5:4556. https://doi.org/10.1038/ncomms5556

Russell FD, Skepper JN, Davenport AP (1997) Detection of endothelin receptors in human coronary artery vascular smooth muscle cells but not endothelial cells by using electron microscope autoradiography. J Cardiovasc Pharmacol 29:820–826

Sakurai T, Yanagisawa M, Takuwa Y, Miyazaki H, Kimura S, Goto K, Masaki T (1990) Cloning of a cDNA encoding a non-isopeptide-selective subtype of the endothelin receptor. Nature 348:732–735. https://doi.org/10.1038/348732a0

Sassone-Corsi P, Visvader J, Ferland L, Mellon PL, Verma IM (1988) Induction of proto-oncogene fos transcription through the adenylate cyclase pathway: characterization of a cAMP-responsive element. Genes Dev 2:1529–1538

Schaeffer C, Thomassin L, Rochette L, Connat JL (2003a) Apoptosis induced in vascular smooth muscle cells by oxidative stress is partly prevented by pretreatment with CGRP. Ann N Y Acad Sci 1010:733–737

Schaeffer C, Vandroux D, Thomassin L, Athias P, Rochette L, Connat JL (2003b) Calcitonin gene-related peptide partly protects cultured smooth muscle cells from apoptosis induced by an oxidative stress via activation of ERK1/2 MAPK. Biochim Biophys Acta 1643:65–73

Schafer C, Steffen H, Printz H, Goke B (1994) Effects of synthetic cyclic AMP analogs on amylase exocytosis from rat pancreatic acini. Can J Physiol Pharmacol 72:1138–1147

Scherer EQ, Herzog M, Wangemann P (2002) Endothelin-1-induced vasospasms of spiral modiolar artery are mediated by rho-kinase-induced Ca(2+) sensitization of contractile apparatus and reversed by calcitonin gene-related peptide. Stroke 33:2965–2971

Schlesinger DH, Goldstein G, Niall HD (1975) The complete amino acid sequence of ubiquitin, an adenylate cyclase stimulating polypeptide probably universal in living cells. Biochemistry 14:2214–2218

Seifert H, Sawchenko P, Chesnut J, Rivier J, Vale W, Pandol SJ (1985) Receptor for calcitonin gene-related peptide: binding to exocrine pancreas mediates biological actions. Am J Physiol 249:G147–G151. https://doi.org/10.1152/ajpgi.1985.249.1.G147

Shenoy SK, McDonald PH, Kohout TA, Lefkowitz RJ (2001) Regulation of receptor fate by ubiquitination of activated beta 2-adrenergic receptor and beta-arrestin. Science 294:1307–1313. https://doi.org/10.1126/science.1063866

Shih SC, Katzmann DJ, Schnell JD, Sutanto M, Emr SD, Hicke L (2002) Epsins and Vps27p/Hrs contain ubiquitin-binding domains that function in receptor endocytosis. Nat Cell Biol 4:389–393. https://doi.org/10.1038/ncb790

Simonetti M, Giniatullin R, Fabbretti E (2008) Mechanisms mediating the enhanced gene transcription of P2X3 receptor by calcitonin gene-related peptide in trigeminal sensory neurons. J Biol Chem 283:18743–18752. https://doi.org/10.1074/jbc.M800296200

Smillie SJ, Brain SD (2011) Calcitonin gene-related peptide (CGRP) and its role in hypertension. Neuropeptides 45:93–104. https://doi.org/10.1016/j.npep.2010.12.002

Sposini S, Hanyaloglu AC (2017) Spatial encryption of G protein-coupled receptor signaling in endosomes; mechanisms and applications. Biochem Pharmacol 143:1–9. https://doi.org/10.1016/j.bcp.2017.04.028

Storer RJ, Akerman S, Goadsby PJ (2004) Calcitonin gene-related peptide (CGRP) modulates nociceptive trigeminovascular transmission in the cat. Br J Pharmacol 142:1171–1181. https://doi.org/10.1038/sj.bjp.0705807

Story GM et al (2003) ANKTM1, a TRP-like channel expressed in nociceptive neurons, is activated by cold temperatures. Cell 112:819–829

Szymanska E, Budick-Harmelin N, Miaczynska M (2018) Endosomal "sort" of signaling control: the role of ESCRT machinery in regulation of receptor-mediated signaling pathways. Semin Cell Dev Biol 74:11–20. https://doi.org/10.1016/j.semcdb.2017.08.012

Takami K, Hashimoto K, Uchida S, Tohyama M, Yoshida H (1986) Effect of calcitonin gene-related peptide on the cyclic AMP level of isolated mouse diaphragm. Jpn J Pharmacol 42:345–350

Takamori M, Yoshikawa H (1989) Effect of calcitonin gene-related peptide on skeletal muscle via specific binding site and G protein. J Neurol Sci 90:99–109

Theoharides TC, Donelan J, Kandere-Grzybowska K, Konstantinidou A (2005) The role of mast cells in migraine pathophysiology. Brain Res Brain Res Rev 49:65–76. https://doi.org/10.1016/j.brainresrev.2004.11.006

Thomsen ARB et al (2016) GPCR-G protein-beta-arrestin super-complex mediates sustained G protein signaling. Cell 166:907–919. https://doi.org/10.1016/j.cell.2016.07.004

Trevisani M et al (2007) 4-Hydroxynonenal, an endogenous aldehyde, causes pain and neurogenic inflammation through activation of the irritant receptor TRPA1. Proc Natl Acad Sci U S A 104:13519–13524. https://doi.org/10.1073/pnas.0705923104

Tsvetanova NG, Irannejad R, von Zastrow M (2015) G protein-coupled receptor (GPCR) signaling via heterotrimeric G proteins from endosomes. J Biol Chem 290:6689–6696. https://doi.org/10.1074/jbc.R114.617951

Urbe S et al (2003) The UIM domain of Hrs couples receptor sorting to vesicle formation. J Cell Sci 116:4169–4179. https://doi.org/10.1242/jcs.00723

Van Dyke RW (2004) Heterotrimeric G protein subunits are located on rat liver endosomes. BMC Physiol 4:1. https://doi.org/10.1186/1472-6793-4-1

Van Valen F, Piechot G, Jurgens H (1990) Calcitonin gene-related peptide (CGRP) receptors are linked to cyclic adenosine monophosphate production in SK-N-MC human neuroblastoma cells. Neurosci Lett 119:195–198

Vause CV, Durham PL (2009) CGRP stimulation of iNOS and NO release from trigeminal ganglion glial cells involves mitogen-activated protein kinase pathways. J Neurochem 110:811–821. https://doi.org/10.1111/j.1471-4159.2009.06154.x

Vause CV, Durham PL (2010) Calcitonin gene-related peptide differentially regulates gene and protein expression in trigeminal glia cells: findings from array analysis. Neurosci Lett 473:163–167. https://doi.org/10.1016/j.neulet.2010.01.074

Walker CS et al (2015) A second trigeminal CGRP receptor: function and expression of the AMY1 receptor. Ann Clin Transl Neurol 2:595–608. https://doi.org/10.1002/acn3.197

Walker CS, Raddant AC, Woolley MJ, Russo AF, Hay DL (2017) CGRP receptor antagonist activity of olcegepant depends on the signalling pathway measured. Cephalalgia:333102417691762. https://doi.org/10.1177/0333102417691762

Wendel-Wellner M, Noll T, Konig P, Schmeck J, Koch T, Kummer W (2002) Cellular localization of the endothelin receptor subtypes ET(A) and ET(B) in the rat heart and their differential expression in coronary arteries, veins, and capillaries. Histochem Cell Biol 118:361–369. https://doi.org/10.1007/s00418-002-0457-4

Weston C et al (2016) Receptor activity-modifying protein-directed G protein signaling specificity for the calcitonin gene-related peptide family of receptors. J Biol Chem 291:21925–21944. https://doi.org/10.1074/jbc.M116.751362

Wilkinson KD (2005) The discovery of ubiquitin-dependent proteolysis. Proc Natl Acad Sci U S A 102:15280–15282. https://doi.org/10.1073/pnas.0504842102

Wu Y et al (2015) Lentivirus mediated over expression of CGRP inhibited oxidative stress in Schwann cell line. Neurosci Lett 598:52–58. https://doi.org/10.1016/j.neulet.2015.05.009

Yanagisawa M et al (1988) A novel potent vasoconstrictor peptide produced by vascular endothelial cells. Nature 332:411–415. https://doi.org/10.1038/332411a0

Yarwood RE et al (2017) Endosomal signaling of the receptor for calcitonin gene-related peptide mediates pain transmission. Proc Natl Acad Sci U S A 114:12309–12314. https://doi.org/10.1073/pnas.1706656114

Zhang J, Ferguson SS, Barak LS, Menard L, Caron MG (1996) Dynamin and beta-arrestin reveal distinct mechanisms for G protein-coupled receptor internalization. J Biol Chem 271:18302–18305

Zhang Z, Winborn CS, Marquez de Prado B, Russo AF (2007) Sensitization of calcitonin gene-related peptide receptors by receptor activity-modifying protein-1 in the trigeminal ganglion. J Neurosci 27:2693–2703. https://doi.org/10.1523/JNEUROSCI.4542-06.2007

Zhou Z, Hu CP, Wang CJ, Li TT, Peng J, Li YJ (2010) Calcitonin gene-related peptide inhibits angiotensin II-induced endothelial progenitor cells senescence through up-regulation of klotho expression. Atherosclerosis 213:92–101. https://doi.org/10.1016/j.atherosclerosis.2010.08.050

Pathways of CGRP Release from Primary Sensory Neurons

Francesco De Logu, Romina Nassini, Lorenzo Landini, and Pierangelo Geppetti

Contents

F. De Logu · R. Nassini · L. Landini · P. Geppetti (✉)
Department of Health Sciences, Section of Clinical Pharmacology and Oncology, Headache Center, University of Florence, Florence, Italy
e-mail: geppetti@unifi.it

© Springer International Publishing AG, part of Springer Nature 2018
S. D. Brain, P. Geppetti (eds.), *Calcitonin Gene-Related Peptide (CGRP) Mechanisms*,
Handbook of Experimental Pharmacology 255, https://doi.org/10.1007/164_2018_145

Abstract

The benefit reported in a variety of clinical trials by a series of small molecule antagonists for the calcitonin gene-related peptide (CGRP) receptor, or four monoclonal antibodies against the neuropeptide or its receptor, has underscored the release of CGRP from terminals of primary sensory neurons, including trigeminal neurons, as one of the major mechanisms of migraine headaches. A large variety of excitatory ion channels and receptors have been reported to elicit CGRP release, thus proposing these agonists as migraine-provoking agents. On the other side, activators of inhibitory channels and receptors may be regarded as potential antimigraine agents. The knowledge of the intracellular pathways underlying the exocytotic process that results in CGRP secretion or its inhibition is, therefore, of importance for understanding how migraine pain originates and how to treat the disease.

Keywords

CGRP · Migraine · Neurogenic inflammation · Primary Sensory Neurons

1 Sources of CGRP Release

1.1 Primary Sensory Neurons

The seminal findings that increased levels of calcitonin gene-related peptide (CGRP), but not of other neuropeptides, were found in plasma from the jugular veins that collect intra- and extracranial blood during migraine or cluster headache attacks (Fanciullacci et al. 1995; Goadsby et al. 1990) have underlined the role of CGRP in the mechanisms of primary headaches (Edvinsson and Warfvinge 2017). Main anatomical structures where the precursor protein of the pre-pro-CGRP is produced and processed are a subpopulation of primary sensory neurons and a subpopulation of intrinsic neurons of the gut (Russell et al. 2014). The α-CGRP is the final product of the CGRP precursor that, from the soma of pseudo-unipolar peptidergic sensory neurons, is transported to their central and peripheral endings. It is stored at both sites in large-core vesicles. These neurons include dorsal root ganglion (DRG), vagal ganglion (VG), and trigeminal ganglion (TG) neurons. CGRP-expressing sensory neurons encompass heterogeneous subpopulations of DRG neurons in terms of anatomy, electrophysiology, neurochemistry, and function. Thus, CGRP has been found in most of the small-sized and some of the intermediate-sized neurons (Gibson et al. 1984), with myelinated slow-conduction C fibers or thinly myelinated Aδ fibers with a higher conduction velocity.

The tachykinins, substance P (SP) and neurokinin A (NKA), are present as peptide neurotransmitters in a proportion of small-sized DRG/VG/TG neurons and are often colocalized with CGRP (Gibson et al. 1984). Usually, SP and NKA are co-released with CGRP from peripheral and central endings of DRG/VG/TG neurons that express the capsaicin-sensitive member of the transient receptor potential (TRP) family of channels, the vanilloid 1 subtype (TRPV1). Notably, in human peripheral tissues innervated by the trigeminal nerve, the release of CGRP, but not that of SP, has been reported (Geppetti et al. 1992).

1.2 Intrinsic Gut Neurons

While α-CGRP expression is confined to DRG/VG/TG primary sensory neurons, the β-isoform, which differs from α-CGRP by only three amino acids in humans, is mainly produced by intrinsic neurons of the enteric nervous system (Russell et al. 2014). Although the dichotomy in the localization of the two CGRP isoforms has been challenged (Li et al. 2009), β-CGRP is usually found in the intestines, with concentrations up to seven times more than α-CGRP (Mulderry et al. 1988). β-CGRP expressed by intrinsic neurons of the gut, which are TRPV1-negative and are therefore capsaicin-insensitive, is spared by the depleting action that results from exposure to high doses of capsaicin (Mulderry et al. 1988). Nevertheless, possible co-expression and co-release of the two isoforms in different areas of the peripheral nervous system, depending on specific circumstances, cannot be excluded. It should be further noted that CGRP-positive fibers originating from extrinsic TRPV1-positive spinal sensory neurons are present in mammals (Tan et al. 2010). In principle, both CGRP isoforms may be released from the gastrointestinal tissues, by non-specific stimuli, whereas capsaicin should solely release α-CGRP.

1.3 Central Neurons

Although CGRP and CGRP receptors have been found in a variety of brain regions (Edvinsson and Warfvinge 2017), their function in structures located inside the blood-brain barrier (BBB) is still the object of much uncertainty. Modulation of CGRP release by a repertoire of ion channels and receptors from peripheral terminals of nociceptors has been reproduced in release experiments from central fibers terminating within the dorsal spinal cord. However, conclusive evidence that such release is associated with nociceptive transmission is lacking. Additional localization of CGRP has been found in neurons of the ventral horns, apparently identified as motoneurons, whereas no CGRP-positive neurons have been reported in dorsal horn (Gibson et al. 1984).

1.4 Non-neuronal Cells

CGRP may be expressed in non-neuronal cells. β-CGRP mRNA and, to a lesser extent, α-CGRP (Hou et al. 2011) have been found in human and rodent-cultured keratinocytes. Both α-CGRP and β-CGRP have also been reported in endothelial progenitor cells that accumulate in damaged endothelium to repair injury and influence vascular remodeling (Zhao et al. 2007). CGRP expression seems to be more abundant in early rather than late endothelial progenitor cells (Fang et al.

2011). There is also some evidence that CGRP is produced by several types of immune cells, including lymphocytes, and peripheral blood mononuclear cells, including activated but not resting B lymphocytes (Bracci-Laudiero et al. 2002), monocytes, and macrophage-activated adipocytes (Linscheid et al. 2004). However, it should be underlined that the proof that CGRP is released from these non-neuronal cells, thus exerting a function, is still lacking.

2   Excitatory Receptors in Primary Sensory Neurons

Although various neuronal and non-neuronal cells may express, and potentially release, CGRP, the neuropeptide important in migraine should derive from primary sensory neurons of the trigeminal ganglion and, possibly, from the upper cervical DRGs. In order to elicit a biologically meaningful discharge of CGRP, or for its attenuation, the neurons should be enriched by excitatory or inhibitory receptors/channels, respectively (Fig. 1). In this paper we will focus on the ones that appear to have a major pathophysiological or therapeutic relevance for migraine.

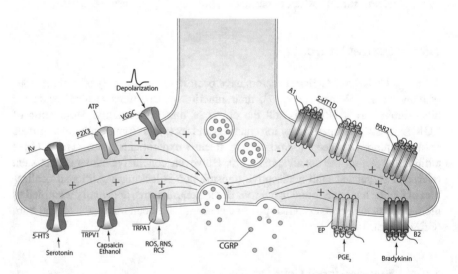

Fig. 1 Schematic representation of peptidergic, CGRP-containing peripheral nerve terminal. Excitatory (+) or inhibitory (−) pathways promote or attenuate, respectively, CGRP release. *Kv* potassium channel, *ATP* adenosine triphosphate, *P2X3* ligand-gated ion channel receptor, *VGSC* voltage-gated sodium channel, *A1* adenosine receptor 1, *5-HT$_{1D}$* 5-hydroxytryptamine (serotonin) receptor 1D subtype, *PAR2* protease-activated receptor 2, *B2* bradykinin 2 receptor, *PGE$_2$* prostaglandin E$_2$, *EP* prostaglandin E receptor, *CGRP* calcitonin gene-related peptide, *ROS* reactive oxygen species, *RNS* reactive nitrogen species, *RCS* reactive carbonyl species, *TRPA1* transient receptor potential ankyrin 1, *TRPV1* transient receptor potential vanilloid 1, *5-HT$_3$* 5-hydroxytryptamine (serotonin) receptor 3 subtype

2.1 Receptors for ATP

Receptors for adenosine triphosphate (ATP) are classified in ligand-gated P2X and metabotropic P2Y receptors, the first group encompassing ion channel receptors and the second G-protein-coupled receptors. P2X2 and P2X3 have been shown in TG neurons (Staikopoulos et al. 2007), where P2X receptors may sensitize sensory neuron functions, including the release of CGRP. P2X3 receptor stimulation caused sensitization of TG neurons, such that a subthreshold amount of KCl was sufficient to increase intracellular Ca^{2+} levels and CGRP secretion (Masterson and Durham 2010). Activation of P2Y receptors in rabbit esophagus mucosa potentiated HCl-evoked release of CGRP (Ma et al. 2011). However, minimal CGRP release was reported following ATP stimulation of P2Y receptors of isolated adult rat DRG neurons (Sanada et al. 2002). Furthermore, ATP did not cause any meaningful release of CGRP from sensory nerve terminals in areas of relevance for migraine mechanism, such as the rat dura mater (Zimmermann et al. 2002). ATP via PY2 receptor activation solely facilitated low pH-evoked release (Zimmermann et al. 2002). As for the action of CGRP on P2X-expressing neurons, it has been reported that, while a proportion of CGRP-binding TG neurons are P2X3-immunopositive, CGRP does not acutely affect the ATP receptor functioning (Fabbretti et al. 2006). Thus, while an indirect tissue-dependent action of ATP via PY2 receptors may favor the release of sensory CGRP, a direct action of ATP in the neurosecretory process is unlikely.

2.2 PAR Receptors

Protease-activated receptor 2 (PAR2) belongs to the PAR family, which encompasses four members. Although PAR1 is expressed by nociceptors and elicits neuropeptide-mediated responses (de Garavilla et al. 2001), more information has been obtained regarding PAR2 that coexists with TRPV1 in peptidergic spinal afferent neurons. Stimulation of PAR2 receptors has been shown to release SP and CGRP, thus producing neurogenic inflammatory responses (Steinhoff et al. 2000). Exposure to PAR2 agonists was found to enhance TRPV1 expression in sensory neurons and to potentiate capsaicin-evoked and TRPV1-dependent currents or Ca^{2+} response in isolated DRG neurons and CGRP release from dorsal spinal cord (Amadesi et al. 2004). In addition, PAR2 activation by stimulating the release of CGRP and tachykinins from capsaicin-sensitive sensory neurons was found to trigger the cytoprotective secretion of gastric mucus in rats (Kawabata et al. 2001). More importantly, supernatants from cultured human pancreatic cancer tissues induced CGRP release from DRG neurons, in a manner that was attenuated by selective PAR2 antagonists (Zhu et al. 2017).

2.3   Serotonin Receptors

Of the large number of serotonin receptors, only $5\text{-HT}_{1B/D}$ and 5-HT_2 are involved in migraine mechanism as selective receptor agonists or partial agonists provide either acute or chronic benefit in migraine, respectively. The ligand-gated ionotropic 5-HT3 receptor is the only serotonin receptor with excitatory activity that, upon activation, has been associated with the release of CGRP (Hua and Yaksh 1993; Tramontana et al. 1993). However, there is no evidence that endogenously released serotonin acts on the 5-HT_3 receptor to release CGRP and, thus, contributes to the migraine mechanism.

2.4   Bradykinin and Prostaglandin Receptors

The proinflammatory and proalgesic peptide, bradykinin, has been reported to release SP and CGRP in vivo and from different isolated preparations in vitro, including the guinea pig heart and the rat dorsal spinal cord via activation of B2 receptors (Figini et al. 1995; Geppetti et al. 1988b). Although expression of B2 receptors has been documented in primary sensory neurons (Steranka et al. 1988), release of sensory neuropeptides does not seem to depend on a direct B2-mediated action of bradykinin on nerve terminals. Attenuation of bradykinin-evoked release of CGRP from isolated guinea pig atria (Geppetti et al. 1990). The reference is correct and rat spinal cord (Andreeva and Rang 1993) by indomethacin indicates the involvement of prostanoids. This conclusion was strengthened by the observation in the isolated guinea pig heart that both arachidonic acid- and bradykinin-evoked CGRP release were abated or markedly attenuated by indomethacin, respectively (Geppetti et al. 1991).

The proposal that prostanoids play a major role on the release of CGRP evoked by arachidonic acid (Geppetti et al. 1990) has been confirmed and amplified by subsequent studies. There are indications that prostaglandins are the final common pathway of several proinflammatory mediators to promote CGRP release, as cyclo-oxygenase inhibition attenuated the neuropeptide outflow from a rat skin preparation evoked by a combination of bradykinin, serotonin, and histamine (Averbeck and Reeh 2001). Of interest for migraine mechanisms and treatment, CGRP release evoked by bradykinin from the intracranial vessels of the guinea pig, similar to results obtained in the heart, was abolished by pretreatment with indomethacin (Geppetti et al. 1990). To strengthen the role of endogenously released prostanoids in CGRP release, direct exposure of cultured DRG neurons to PGE_2 and PGI_2 evoked CGRP release (Hingtgen et al. 1995). Efficacy of nonsteroidal anti-inflammatory drugs (NSAIDs) to abort migraine attacks may depend on their ability to inhibit prostaglandin synthesis and the ensuing release of the pro-migraine neuropeptide, CGRP.

3 Excitatory Channels in Trigeminal Primary Sensory Neurons

3.1 Sodium Channels

Nine different voltage-gated sodium channel (VGSC) isoforms (Nav1.1–Nav1.9) that share a common overall structural motif, but with different amino acid sequences, have been described (Catterall 2000). Genetic and pharmacological findings in experimental animals and humans have implicated some of them in the mechanism of different types of pain, especially the tetrodotoxin (TTX)-sensitive Nav1.3 and Nav1.7 and the TTX-resistant Nav1.8 and Nav1.9 (Catterall 2000). Antidromic invasion of terminal fibers of nociceptors by propagated action potential (Bayliss 1901) has been hypothesized to account for the activation paradigm that results in the liberation of a chemical mediator that promotes vasodilatation and sensitization of neighboring sensory fibers (Lewis 1936). There is now conclusive evidence that indicates CGRP as the substance released locally from cutaneous sensory nerve terminals that mediates neurogenic vasodilatation (Sinclair et al. 2010).

Depolarization produced by electrical field stimulation or high K^+ concentrations results in the release of sensory neuropeptides via the opening of TTX-sensitive VGSCs and the ensuing influx of extracellular Ca^{2+}, usually through voltage-gated calcium channels (VGCC) of the N-type (Lundberg et al. 1992). If this is the typical way of releasing SP, NKA and CGRP from central and peripheral terminals of sensory neurons, it is by no means exclusive. For example, the prostaglandin-dependent release pathway activated by bradykinin is in large part TTX-resistant, whereas VGCCs are markedly involved (Geppetti 1993). Furthermore, additional and diverse mechanisms contributing to the release of sensory neuropeptides, particularly that one elicited by the hot spice, capsaicin, which is a selective TRPV1 agonist, are also entirely TTX-resistant (Maggi et al. 1988).

3.2 TRP Channels

TRPV1 belongs to a larger family of ligand-gated cation channels that in mammals encompasses 28 members grouped in 6 families distinguished on the basis of their sequence homology (Nilius and Szallasi 2014). Some of them are expressed in primary sensory neurons, thereby attracting interest for their possible roles in pain mechanisms. These include the vanilloid 2, 3, and 4 subtypes (TRPV2, TPV3, and TRPV4), the melastatin 3 and 8 subtypes (TRPM3 and TRPM8), and the ankyrin 1 subtype (TRPA1). They have been labeled as thermosensors, because of their ability to sense different temperatures from noxious cold (TRPA1) to very noxious heat (TRPV2). However, the coding for efficient detection of temperature appears more complex and integrated, as deletion of the three different channels, TRPV1, TRPM3 and TRPA1, is required to abolish heat sensation in mice (Vandewauw et al. 2018).

More importantly for the present discussion, in the last decade, a number of reports have shown that some TRP channels are particularly sensitive to the redox state of the milieu, the TRPA1 channel being the most sensitive (Mori et al. 2016). In fact, a large series of reactive oxygen (ROS), nitrogen (RNS), and carbonyl (RCS) species target TRPA1 by binding to specific cysteine residues (C415S, C422S, and C622S) of the intracellular domain (Macpherson et al. 2007). The unprecedented list of endogenous molecules that gate TRPA1 includes the ROS, H_2O_2 (Sawada et al. 2008); the RNS, peroxynitrite (Andersson et al. 2015); and the RCS, 4-hydroxynonenal (Trevisani et al. 2007). Nevertheless, seminal studies that have fully revealed the role of nociceptors in releasing proinflammatory neuropeptides from their peripheral terminals, to mediate neurogenic inflammatory responses, have been possible by using capsaicin that selectively targets TRPV1.

Long before the cloning of TRPV1 (Caterina et al. 1997), the understanding of the unique physiopharmacology of the "capsaicin receptor" was clear. Exposure to high capsaicin concentrations via a massive influx of extracellular Ca^{2+} into the nerve terminal elicits two distinct and time-related effects. First, capsaicin excites the nerve terminal, producing the classical burning pain sensation coupled to the release of sensory neuropeptides and the ensuing neurogenic inflammatory responses. Shortly after, probably due to the neurotoxic action of excessive cytoplasmic Ca^{2+} levels gained by prolonged exposure to high capsaicin concentrations, nerve terminals become insensitive to further stimulation by capsaicin or any other stimulus (Bevan and Szolcsanyi 1990). In vivo, systemic administration or topical application of capsaicin produces the same dual effects observed in vitro, which, however, are characterized by a time-dependent recovery (Geppetti et al. 1988a). The so-called capsaicin desensitization has been exploited in the use of topical preparations for the treatment of post-herpetic or HIV-related neuralgias (Katz et al. 2015).

The ability of capsaicin exposure to elicit and attenuate pain sensation parallels that to cause and inhibit neuropeptide release. Thus, tissues containing peptidergic sensory neurons after exposure to capsaicin undergo an initial release of neuropeptides, followed by refractoriness to any further release. An example of such a property has been reported in human tissues containing trigeminal sensory fibers, such as the iris and ciliary body. Capsaicin released CGRP at a first, but not a second, exposure to capsaicin (Geppetti et al. 1992). Notably, capsaicin failed to release SP from this trigeminal human preparation, underlying that CGRP, and not SP, mediates neurogenic inflammation in humans (Geppetti et al. 1992). The observation that telcagepant, one of the first small molecule antagonists of the CGRP receptor, abated the vasodilatation evoked by capsaicin application to the human forearm (Sinclair et al. 2010) demonstrated that CGRP is the compound released from sensory nerves to dilate arterioles and sensitize neighboring terminal fibers (Bayliss 1901; Lewis 1936). It also suggests that beneficial effects in cluster headache or migraine of repeated topical application of capsaicin to areas innervated by the trigeminal nerve is due to capsaicin desensitization (Fusco et al. 2003) and the ensuing reduced ability to release CGRP.

A series of pharmacological and genetic data point to TRPA1 as one of the major transducers that links CGRP release, local inflammation, neuronal sensitization, and pain. Indirect evidence has associated oxidants and reactive molecules with migraine, as indicated by increased nitrate/nitrite levels found in premenopausal women in association with their migraine (Tietjen et al. 2009). More generally, preclinical and clinical findings underscore the role of oxidative stress in migraine (Borkum 2018). However, the oxidant-dependent pathways that lead to oxidant generation that in turn promote the CGRP-dependent migraine pain remain to be investigated.

4   Inhibitory Receptors and Channels in Trigeminal Primary Sensory Neurons

4.1   Adenosine Receptors

There is evidence that drugs that increase the levels of cAMP or cGMP, such as cilostazol or sildenafil, respectively, are detrimental for migraine (Ashina et al. 2017). Thus, it is not surprising that activation of receptors coupled to a G-protein that inhibits the activity of adenylyl cyclase and reduces the intracellular levels of cAMP is beneficial in migraine. A selective agonist for the adenosine receptor 1 (A1), GR79236, inhibited the release of CGRP evoked by superior sagittal sinus stimulation in the cat (Goadsby et al. 1988). Initial enthusiasm for new therapeutic opportunities for migraine by activating such inhibitory receptors was attenuated by cardiovascular adverse effects, such as bradycardia, and the issue whether A1 stimulation may cause vasoconstriction rather than inhibition of neurotransmitter release (Arulmani et al. 2005).

4.2   Serotonin 5-HT$_{1D}$ Receptors

More clinically relevant information has been obtained on the localization and function of the serotonin 5-HT$_{1B}$ and 5-HT$_{1D}$ subtypes, expressed by the vascular smooth muscle and the sensory nerve terminal, respectively. Their association with a Gi protein, which inhibits adenylyl cyclase, indicates that their stimulation causes vasoconstriction and inhibits the release of SP and CGRP, in vascular smooth muscle and sensory nerve terminals, respectively (Moskowitz 1992). Ergot derivatives, such as ergotamine or dihydroergotamine, target 5-HT$_{1B/D}$ receptors but also activate 5-HT$_2$, α-adrenergic, and dopamine receptors, which markedly contribute to their adverse reactions. As triptans selectively stimulate the 5-HT$_{1B/D}$ receptor, migraine amelioration should solely depend on these receptor subtypes. However, there is no final conclusion as to whether resolution of the migraine attack by triptans or ergots is associated with 5-HT_{1B}-mediated vasoconstriction or 5-HT_{1D}-mediated attenuation of CGRP release or both. Positive clinical evidence in migraine treatment with lasmiditan, a ditan with a selective agonistic action on the 5-HT_{1F} subtype (Rubio-Beltran et al. 2018), whose expression is apparently confined to sensory nerve terminals, suggests that inhibition of CGRP release should be the major antimigraine mechanism of triptans/ergots. However, a firm conclusion on this long-lasting debate has not yet been reached.

4.3   Potassium Channels

Opening and closing of inhibitory K^+ channels mediate fine tuning of neuronal responses. A triple cysteine pocket within neuronal M-channel subunits (Kv7.2–7.5) is oxidatively modified by H_2O_2 (Gamper et al. 2006) and nitric oxide (NO) to control the release of CGRP (Ooi et al. 2013). The gaseotransmitter modulator, hydrogen sulfide (H_2S), may activate both the TRPV1 and TRPA1 channels, thereby evoking the release of CGRP (Pozsgai et al. 2012). However, H_2S produces mechanical allodynia and increases neuronal excitability in TG neurons, via an autocrine mechanism underlined by the colocalization of the H_2S-generating enzyme cystathionine beta-synthase (CBS) and H_2S-sensitive Kv1.1 and Kv1.4 channels (Feng et al. 2013).

5   Agents that Provoke Migraine Attacks and Release CGRP

5.1   Nitric Oxide Donors

Headaches can be provoked by occupational exposure to, or treatment with, organic nitrates (Thadani and Rodgers 2006; Trainor and Jones 1966). From these observations, the NO donor, glyceryl trinitrate (GTN), has been proposed and used as a provocation test for migraine attacks (Iversen et al. 1989; Sicuteri et al. 1987). GTN causes a mild, rapidly developing, and short-lived headache in most subjects, including healthy controls, which is followed, mainly in migraineurs, by a markedly delayed headache that fulfills the criteria of a typical migraine attack (Iversen et al. 1989; Sicuteri et al. 1987). Thus, the migraine is temporally dissociated from the immediate and short-lived (<10 min) release of NO (Persson et al. 1994) and the consequent cGMP-dependent (Guo et al. 2008) vasodilatation. Degranulation of meningeal mast cells (Ferrari et al. 2016; Reuter et al. 2001), phosphorylation of extracellular signal-regulated kinase (ERK) in meningeal arteries (Zhang et al. 2013), and delayed meningeal inflammation sustained by induction of NO synthase and prolonged NO generation have been proposed to explain the delayed GTN-evoked headache. The ability of NO generated by GTN to release CGRP

(Ramachandran et al. 2014; Strecker et al. 2002) has been also considered. However, unlike the delayed migraine headache, CGRP release after exposure to GTN/NO is rapid. Thus, further mechanisms must be explored to understand the timing and mechanism of migraine evoked by NO donors.

5.2   Ethanol and TRPV1

It is known that migraineurs report headaches after ingestion of alcoholic beverages more easily than healthy controls and that after a considerable delay they may experience real migraine attacks (Kelman 2007). Relatively elevated concentrations of ethanol sensitize TRPV1 by lowering the normal temperature for channel activation by 8°C (from 43 to 35°C), thus allowing normal body temperature to gate the channel (Trevisani et al. 2002). This activating mechanism elicits a Ca^{2+}-dependent secretion of CGRP from peripheral terminals of DRG neurons (Trevisani et al. 2002). By TRPV1 targeting and the ensuing CGRP release, ethanol produces dilatation of meningeal arteries of guinea pigs (Nicoletti et al. 2008). However, there is a temporal dissociation between the rapid and almost immediate TRPV1-dependent and CGRP-mediated vasodilatation and the much more delayed development of migraine attacks. *The International Classification of Headache Disorders, 3rd edition* (IHS 2018) distinguishes an immediate alcohol-induced headache, which develops within 3 h and resolves within 72 h from alcohol ingestion, and the delayed alcohol-induced headache (previously defined hangover headache), which develops within 5–12 h after alcohol ingestion and resolves within 72 h of onset. As after 12 h from ingestion no ethanol is found in plasma, it seems unlikely that TRPV1 and CGRP are the sole contributors of alcohol-related headaches.

5.3   TRPA1 Agonists

Umbellulone is a major constituent of the *Umbellularia californica*, also known as the headache tree, because seasonal exposure to the scent of the plant provokes severe headaches, including attacks of cluster headache (Benemei et al. 2010) in susceptible individuals. Umbellulone may quickly react with thiols of cysteine residues of TRPA1, thereby activating the trigeminovascular system to evoke a CGRP-dependent meningeal vasodilatation (Nassini et al. 2012), indicating that channel targeting may be the underlying mechanism by which the scent of *Umbellularia californica* evokes migraine attacks. Acrolein is contained in combustion exhausts or can be produced endogenously following peroxidation of plasma protein phospholipids. Its ability to target TRPA1 (Bautista et al. 2006) has also been shown in TG neurons where it can dilate meningeal vessels via a CGRP-mediated mechanism (Kunkler et al. 2011) (Fig. 2).

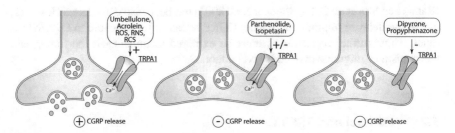

Fig. 2 Schematic representation of the different modes of stimulating or inhibiting CGRP release from peptidergic primary sensory neurons by drugs that target TRPA1, in relation to their ability to provoke or ameliorate migraine, respectively. Umbellulone, which provokes migraine, acrolein, ROS, RNS and RCS are full TRPA1 agonists that target TRPA1 and release CGRP. The prophylactic antimigraine agents, parthenolide and isopetasin, behave as partial TRPA1 agonists that desensitize the channel and the nerve terminal, thus attenuating CGRP release. Dipyrone and propyphenazone, effective drugs for the acute migraine treatment, are selective TRPA1 antagonists

6   Agents that Ameliorate Migraine Attacks and Inhibit CGRP Release

6.1   Nonsteroidal Anti-inflammatory Drugs

We have already addressed the role of triptans that attenuate the release of CGRP and the resultant proinflammatory and proalgesic effects by activating inhibitory 5-HT$_{1D}$ receptors expressed by perivascular nerve terminals. NSAIDs are additional drugs commonly used for the acute treatment of migraine attacks. Their analgesic action relies on the blockade of cyclooxygenases, thereby inhibiting the production of proinflammatory and proalgesic prostaglandins. The ability of prostaglandins (see above) to release CGRP from terminals of peptidergic nociceptors and the ability of NSAIDs to attenuate arachidonic acid or bradykinin-evoked CGRP release (Geppetti 1993) support the view that at least part of the antimigraine action of NSAIDs is due to their ability to attenuate the prostaglandin-evoked (Hingtgen et al. 1995) release of the pro-migraine neuropeptide.

6.2   Herbal Medicines

Herbal medicines have been used for centuries for alleviating pain and headaches. From popular medicine, some of these preparations have gained a more robust position in the therapeutic antimigraine armamentarium because of positive clinical trials and recent acquisitions of their mechanisms of action. Petasin and isopetasin (Avula et al. 2012), contained in butterbur [*Petasites hybridus (L.) Gaertn.*], are considered responsible for the antimigraine effects of the herbal extract. In fact, a preparation containing standardized amounts (minimum 15%, corresponding to 7.5 mg) of petasin/isopetasin (Avula et al. 2012; Danesch and Rittinghausen 2003) showed a beneficial action in migraine prevention (Grossmann and Schmidramsl 2000; Lipton et al. 2004; Pothmann and Danesch 2005). These studies led to the indication by the American Headache Society guidelines with a level A recommendation of butterbur for migraine prophylaxis (Holland et al. 2012). Several hypotheses, including antileukotriene or antimuscarinic activity (Ko et al. 2001; Thomet et al. 2001), have been proposed to explain the antimigraine action of petasin/isopetasin. Recently, by considering that petasin and its cross-conjugated isomer, isopetasin, are eremophilane sesquiterpene esters of petasol and angelic acid, which contain electrophilic double bonds and can potentially interact with bionucleophiles, it has been hypothesized that they could interact with TRPA1 (Benemei et al. 2017).

The observation that isopetasin targets TRPA1 to evoke Ca^{2+} response in TG neurons and to release CGRP from central terminals of TRPA1-expressing neurons (Benemei et al. 2017) was, however, conflicting with the beneficial action of butterbur in migraine. However, the excitatory action of isopetasin on sensory neurons is weak, as it behaves as a TRPA1 partial agonist. In addition, after in vitro or in vivo exposure to isopetasin, the TRPA1 channel and the TRPA1-expressing trigeminal neurons undergo concentration- and dose-dependent desensitization. Thus, after an initial moderate excitatory action, exposure to isopetasin results in prolonged inhibition of nociceptive responses and CGRP release from TRPA1-expressing neurons (Benemei et al. 2017). A similar moderate excitatory action followed by a prolonged desensitizing effect has been previously reported by parthenolide, a major constituent of *Tanacetum parthenium* (Materazzi et al. 2013). Preparations containing the herbal extract or parthenolide are marketed for migraine prophylaxis. Thus, attenuation of TRPA1 activity on trigeminal neurons, associated with a reduced CGRP release, may be a common mechanism of both butterbur and *Tanacetum parthenium* to ameliorate migraine (Fig. 2).

6.3   Pyrazolone Derivatives

A randomized double-blind clinical trial (Bigal et al. 2002) supported the pharmaco-epidemiological observation that the pyrazolone derivative, dipyrone (metamizole), is effective for the acute relief of migraine attacks (Ramacciotti et al. 2007). Although it is not available in some countries (particularly the USA and the UK) because of its association with potentially life-threatening blood dyscrasias such as agranulocytosis, dipyrone remains a successful remedy for treating pain and migraine headaches in other countries. In Brazil, a much larger number of patients (almost 60%) treat their migraine attacks with medicines containing dipyrone alone or in combination, compared to patients who use paracetamol (16%), triptans (6%), or NSAIDs (12%) (Chagas et al. 2015).

Dipyrone and the associated pyrazolone derivative, propyphenazone, which is also successfully used for acute migraine treatment, are commonly included in the larger family of NSAIDs. However, previous (Lorenzetti and Ferreira 1985) and more recent reports (Nassini et al. 2015) have argued against this conclusion. While indomethacin was equipotent in reducing edema and hyperalgesia evoked by carrageenan in rats, dipyrone showed a remarkable anti-hyperalgesic action but a poor anti-inflammatory effect (Lorenzetti and Ferreira 1985). More recently, it has been reported that carrageenan-evoked mechanical allodynia was attenuated by dipyrone and propyphenazone, without affecting local prostaglandin E_2 levels (Nassini et al. 2015). As dipyrone and propyphenazone selectively inhibited TRPA1-dependent Ca^{2+} responses and currents and attenuated TRPA1-mediated pain-like responses in models of inflammatory and neuropathic pain (formalin, carrageenan, partial sciatic nerve ligation, and bortezomib), it was proposed that analgesia by dipyrone and propyphenazone is due to their TRPA1 antagonistic effect. Finally, dipyrone and propyphenazone reduced the TRPA1-evoked release of CGRP from primary sensory neurons, a response that may account for the antimigraine action of these drugs (Nassini et al. 2015) (Fig. 2).

7 Conclusion

The release of neurotransmitters is usually initiated by propagated action potentials which invade nerve terminals through the opening VGSC. This event is followed by the opening of the VGCC, which, allowing Ca^{2+} influx, promotes the migration and fusion with the plasma membrane of the vesicles that store neurotransmitters, thus leading to neurosecretion. Regarding trigeminal primary sensory neurons, orthodromically propagated action potentials from the periphery invade central terminals to elicit the CGRP release in the lamina I and II of the dorsal horn of the brain stem inside the BBB. It is therefore unlikely that small molecules (olcegepant or telcagepant) which poorly penetrate the BBB (Hostetler et al. 2013; Tfelt-Hansen and Olesen 2011) exert their antimigraine activity (Ho et al. 2008; Olesen et al. 2004) at such a central site of action. Efficacy in migraine treatment of recently developed anti-CGRP receptor monoclonal antibodies (Goadsby et al. 2017; Silberstein et al. 2017), whose penetration of the BBB is limited to 0.1–0.5%, further points to a peripheral action of these new antimigraine medicines.

The observation that effective migraine treatment is associated with blockade of CGRP receptors outside the BBB implies that migraine mechanism is mediated by CGRP release not from central, but rather from peripheral endings of trigeminal primary sensory neurons. There are two neurophysiological pathways that result in peripheral CGRP release. The first is driven by antidromically propagated action potentials that, in a TTX-sensitive manner, invade very terminal fibers, mostly surrounding meningeal or pericranial arterial vessels, to promote a VGCC-dependent neurosecretion. The second is promoted by the activation of ligand-gated ion channels, including the acid-sensitive TRPV1 and the oxidative stress sensor TRPA1, as well as a number of excitatory receptors expressed on the cell membrane. This second pathway is TTX-insensitive and mostly independent from VGCC. Finally, we underline that we owe most of our current understanding of the subtle mechanisms underlying the outflow of CGRP from TG neurons that have disclosed new avenues for the treatment of migraine to the pioneering studies on antidromic conduction in sensory neurons, which promote neurogenic inflammation and sensitization to pain (Bayliss 1901; Lewis 1936), and to the use of capsaicin (Bevan and Szolcsanyi 1990), the prototypical activator and desensitizing agent of TRPV1, and peptidergic nociceptors.

Acknowledgments We thank the Migraine Research Foundation, New York (Grant 2017), and Ministero dell'Istruzione, dell'Università e della Ricerca (MiUR), Rome (PRIN 201532AHAE_003).

References

Amadesi S, Nie J, Vergnolle N, Cottrell GS, Grady EF, Trevisani M, Manni C, Geppetti P, McRoberts JA, Ennes H, Davis JB, Mayer EA, Bunnett NW (2004) Protease-activated receptor 2 sensitizes the capsaicin receptor transient receptor potential vanilloid receptor 1 to induce hyperalgesia. J Neurosci 24:4300–4312

Andersson DA, Filipovic MR, Gentry C, Eberhardt M, Vastani N, Leffler A, Reeh P, Bevan S (2015) Streptozotocin stimulates the Ion Channel TRPA1 directly: involvement of peroxynitrite. J Biol Chem 290:15185–15196

Andreeva L, Rang HP (1993) Effect of bradykinin and prostaglandins on the release of calcitonin gene-related peptide-like immunoreactivity from the rat spinal cord in vitro. Br J Pharmacol 108:185–190

Arulmani U, Heiligers JP, Centurion D, Garrelds IM, Villalon CM, Saxena PR (2005) Lack of effect of the adenosine A1 receptor agonist, GR79236, on capsaicin-induced CGRP release in anaesthetized pigs. Cephalalgia 25:1082–1090

Ashina M, Hansen JM, Á Dunga BO, Olesen J (2017) Human models of migraine – short-term pain for long-term gain. Nat Rev Neurol 13:713–724

Averbeck B, Reeh PW (2001) Interactions of inflammatory mediators stimulating release of calcitonin gene-related peptide, substance P and prostaglandin E(2) from isolated rat skin. Neuropharmacology 40:416–423

Avula B, Wang YH, Wang M, Smillie TJ, Khan IA (2012) Simultaneous determination of sesquiterpenes and pyrrolizidine alkaloids from the rhizomes of Petasites hybridus (L.) G.M. et Sch. and dietary supplements using UPLC-UV and HPLC-TOF-MS methods. J Pharm Biomed Anal 70:53–63

Bautista DM, Jordt SE, Nikai T, Tsuruda PR, Read AJ, Poblete J, Yamoah EN, Basbaum AI, Julius D (2006) TRPA1 mediates the inflammatory actions of environmental irritants and proalgesic agents. Cell 124:1269–1282

Bayliss WM (1901) On the origin from the spinal cord of the vaso-dilator fibres of the hind-limb, and on the nature of these fibres. J Physiol 26:173–209

Benemei S, Appendino G, Geppetti P (2010) Pleasant natural scent with unpleasant effects: cluster headache-like attacks triggered by Umbellularia californica. Cephalalgia 30:744–746

Benemei S, De Logu F, Li Puma S, Marone IM, Coppi E, Ugolini F, Liedtke W, Pollastro F, Appendino G, Geppetti P, Materazzi S, Nassini R (2017) The anti-migraine component of butterbur extracts, isopetasin, desensitizes peptidergic nociceptors by acting on TRPA1 cation channel. Br J Pharmacol 174:2897–2911

Bevan S, Szolcsanyi J (1990) Sensory neuron-specific actions of capsaicin: mechanisms and applications. Trends Pharmacol Sci 11:330–333

Bigal ME, Bordini CA, Tepper SJ, Speciali JG (2002) Intravenous dipyrone in the acute treatment of migraine without aura and migraine with aura: a randomized, double blind, placebo controlled study. Headache 42:862–871

Borkum JM (2018) The migraine attack as a homeostatic, neuroprotective response to brain oxidative stress: preliminary evidence for a theory. Headache 58:118–135

Bracci-Laudiero L, Aloe L, Buanne P, Finn A, Stenfors C, Vigneti E, Theodorsson E, Lundeberg T (2002) NGF modulates CGRP synthesis in human B-lymphocytes: a possible anti-inflammatory action of NGF? J Neuroimmunol 123:58–65

Caterina MJ, Schumacher MA, Tominaga M, Rosen TA, Levine JD, Julius D (1997) The capsaicin receptor: a heat-activated ion channel in the pain pathway. Nature 389:816–824

Catterall WA (2000) From ionic currents to molecular mechanisms. Neuron 26:13–25

Chagas OFP, Eckeli FD, Bigal ME, da Silva MOA, Speciali JG (2015) Study of the use of analgesics by patients with headache at a specialized outpatient clinic (ACEF). Arq Neuropsiquiatr 73:586–592

Danesch U, Rittinghausen R (2003) Safety of a patented special butterbur root extract for migraine prevention. Headache 43:76–78

de Garavilla L, Vergnolle N, Young SH, Ennes H, Steinhoff M, Ossovskaya VS, D'Andrea MR, Mayer EA, Wallace JL, Hollenberg MD, Andrade-Gordon P, Bunnett NW (2001) Agonists of proteinase-activated receptor 1 induce plasma extravasation by a neurogenic mechanism. Br J Pharmacol 133:975–987

Edvinsson L, Warfvinge K (2017) Recognizing the role of CGRP and CGRP receptors in migraine and its treatment. Cephalalgia. https://doi.org/10.1177/0333102417736900

Fabbretti E, D'Arco M, Fabbro A, Simonetti M, Nistri A, Giniatullin R (2006) Delayed upregulation of ATP P2X3 receptors of trigeminal sensory neurons by calcitonin gene-related peptide. J Neurosci 26:6163–6171

Fanciullacci M, Alessandri M, Figini M, Geppetti P, Michelacci S (1995) Increase in plasma calcitonin gene-related peptide from the extracerebral circulation during nitroglycerin-induced cluster headache attack. Pain 60:119–123

Fang L, Chen MF, Xiao ZL, Liu Y, Yu GL, Chen XB, Xie XM (2011) Calcitonin gene-related peptide released from endothelial progenitor cells inhibits the proliferation of rat vascular smooth muscle cells induced by angiotensin II. Mol Cell Biochem 355:99–108

Feng X, Zhou YL, Meng X, Qi FH, Chen W, Jiang X, Xu GY (2013) Hydrogen sulfide increases excitability through suppression of sustained potassium channel currents of rat trigeminal ganglion neurons. Mol Pain 9:4

Ferrari LF, Levine JD, Green PG (2016) Mechanisms mediating nitroglycerin-induced delayed-onset hyperalgesia in the rat. Neuroscience 317:121–129

Figini M, Javdan P, Cioncolini F, Geppetti P (1995) Involvement of tachykinins in plasma extravasation induced by bradykinin and low pH mediumin the Guinea-pig conjunctiva. Br J Pharmacol 115:128

Fusco BM, Barzoi G, Agrò F (2003) Repeated intranasal capsaicin applications to treat chronic migraine. Br J Anaesth 90:812

Gamper N, Zaika O, Li Y, Martin P, Hernandez CC, Perez MR, Wang AY, Jaffe DB, Shapiro MS (2006) Oxidative modification of M-type K+ channels as a mechanism of cytoprotective neuronal silencing. EMBO J 25:4996–5004

Geppetti P (1993) Sensory neuropeptide release by bradykinin: mechanisms and pathophysiological implications. Regul Pept 47:1–23

Geppetti P, Fusco BM, Marabini S, Maggi CA, Fanciullacci M, Sicuteri F (1988a) Secretion, pain and sneezing induced by the application of capsaicin to the nasal mucosa in man. Br J Pharmacol 93:509–514

Geppetti P, Maggi CA, Perretti F, Frilli S, Manzini S (1988b) Simultaneous release by bradykinin of substance P- and calcitonin gene-related peptide immunoreactivities from capsaicin-sensitive structures in guinea-pig heart. Br J Pharmacol 94:288–290

Geppetti P, Tramontana M, Santicioli P, Blanco ED, Giuliani S, Maggi CA (1990) Bradykinin-induced release of calcitonin gene-related peptide from capsaicin-sensitive nerves in Guinea-pig atria: mechanism of action and calcium requirements. Neuroscience 38:687–692

Geppetti P, Del Bianco E, Tramontana M, Vigano T, Folco GC, Maggi CA, Manzini S, Fanciullacci M (1991) Arachidonic acid and bradykinin share a common pathway to release neuropeptide from capsaicin-sensitive sensory nerve fibers of the guinea pig heart. J Pharmacol Exp Ther 259:759–765

Geppetti P, Del Bianco E, Cecconi R, Tramontana M, Romani A, Theodorsson E (1992) Capsaicin releases calcitonin gene-related peptide from the human iris and ciliary body in vitro. Regul Pept 41:83–92

Gibson SJ, Polak JM, Bloom SR, Sabate IM, Mulderry PM, Ghatei MA, McGregor GP, Morrison JF, Kelly JS, Evans RM et al (1984) Calcitonin gene-related peptide immunoreactivity in the spinal cord of man and of eight other species. J Neurosci 4:3101–3111

Goadsby PJ, Edvinsson L, Ekman R (1988) Release of vasoactive peptides in the extracerebral circulation of humans and the cat during activation of the trigeminovascular system. Ann Neurol 23:193–196

Goadsby PJ, Edvinsson L, Ekman R (1990) Vasoactive peptide release in the extracerebral circulation of humans during migraine headache. Ann Neurol 28:183–187

Goadsby PJ, Reuter U, Hallstrom Y, Broessner G, Bonner JH, Zhang F, Sapra S, Picard H, Mikol DD, Lenz RA (2017) A controlled trial of erenumab for episodic migraine. N Engl J Med 377 (22):2123–2132

Grossmann M, Schmidramsl H (2000) An extract of Petasites hybridus is effective in the prophylaxis of migraine. Int J Clin Pharmacol Ther 38:430–435

Guo R, Chen XP, Guo X, Chen L, Li D, Peng J, Li YJ (2008) Evidence for involvement of calcitonin gene-related peptide in nitroglycerin response and association with mitochondrial aldehyde dehydrogenase-2 (ALDH2) Glu504Lys polymorphism. J Am Coll Cardiol 52:953–960

Hingtgen CM, Waite KJ, Vasko MR (1995) Prostaglandins facilitate peptide release from rat sensory neurons by activating the adenosine 3′,5′-cyclic monophosphate transduction cascade. J Neurosci 15:5411–5419

Ho TW, Ferrari MD, Dodick DW, Galet V, Kost J, Fan X, Leibensperger H, Froman S, Assaid C, Lines C, Koppen H, Winner PK (2008) Efficacy and tolerability of MK-0974 (telcagepant), a new oral antagonist of calcitonin gene-related peptide receptor, compared with zolmitriptan for acute migraine: a randomised, placebo-controlled, parallel-treatment trial. Lancet 372:2115–2123

Holland S, Silberstein SD, Freitag F, Dodick DW, Argoff C, Ashman E (2012) Evidence-based guideline update: NSAIDs and other complementary treatments for episodic migraine prevention in adults: report of the Quality Standards Subcommittee of the American Academy of Neurology and the American Headache Society. Neurology 78:1346–1353

Hostetler ED, Joshi AD, Sanabria-Bohorquez S, Fan H, Zeng Z, Purcell M, Gantert L, Riffel K, Williams M, O'Malley S, Miller P, Selnick HG, Gallicchio SN, Bell IM, Salvatore CA, Kane SA, Li CC, Hargreaves RJ, de Groot T, Bormans G, Van Hecken A, Derdelinckx I, de Hoon J, Reynders T, Declercq R, De Lepeleire I, Kennedy WP, Blanchard R, Marcantonio EE, Sur C, Cook JJ, Van Laere K, Evelhoch JL (2013) In vivo quantification of calcitonin gene-related peptide receptor occupancy by telcagepant in rhesus monkey and human brain using the positron emission tomography tracer [11C]MK-4232. J Pharmacol Exp Ther 347:478–486

Hou Q, Barr T, Gee L, Vickers J, Wymer J, Borsani E, Rodella L, Getsios S, Burdo T, Eisenberg E, Guha U, Lavker R, Kessler J, Chittur S, Fiorino D, Rice F, Albrecht P (2011) Keratinocyte expression of calcitonin gene-related peptide beta: implications for neuropathic and inflammatory pain mechanisms. Pain 152:2036–2051

Hua XY, Yaksh TL (1993) Pharmacology of the effects of bradykinin, serotonin, and histamine on the release of calcitonin gene-related peptide from C-fiber terminals in the rat trachea. J Neurosci 13:1947–1953

IHS (2018) The international classification of headache disorders, 3rd edition. Cephalalgia 38:1–211

Iversen HK, Olesen J, Tfelt-Hansen P (1989) Intravenous nitroglycerin as an experimental-model of vascular headache – basic characteristics. Pain 38:17–24

Katz NP, Mou J, Paillard FC, Turnbull B, Trudeau J, Stoker M (2015) Predictors of response in patients with postherpetic neuralgia and HIV-associated neuropathy treated with the 8% capsaicin patch (Qutenza). Clin J Pain 31:859–866

Kawabata A, Kinoshita M, Nishikawa H, Kuroda R, Nishida M, Araki H, Arizono N, Oda Y, Kakehi K (2001) The protease-activated receptor-2 agonist induces gastric mucus secretion and mucosal cytoprotection. J Clin Invest 107:1443–1450

Kelman L (2007) The triggers or precipitants of the acute migraine attack. Cephalalgia 27:394–402

Ko WC, Lei CB, Lin YL, Chen CF (2001) Mechanisms of relaxant action of S-petasin and S-isopetasin, sesquiterpenes of Petasites formosanus, in isolated Guinea pig trachea. Planta Med 67:224–229

Kunkler PE, Ballard CJ, Oxford GS, Hurley JH (2011) TRPA1 receptors mediate environmental irritant-induced meningeal vasodilatation. Pain 152:38–44

Lewis T (1936) Experiments relating to cutaneous hyperalgesia and its spread through somatic nerves. Clin Sci 2:373–421

Li D, Chen B-M, Peng J, Zhang Y-S, Li X-H, Yuan Q, Hu C-P, Deng H-W, Li Y-J (2009) Role of anandamide transporter in regulating calcitonin gene-related peptide production and blood pressure in hypertension. J Hypertens 27:1224–1232

Linscheid P, Seboek D, Schaer DJ, Zulewski H, Keller U, Muller B (2004) Expression and secretion of procalcitonin and calcitonin gene-related peptide by adherent monocytes and by macrophage-activated adipocytes. Crit Care Med 32:1715–1721

Lipton RB, Gobel H, Einhaupl KM, Wilks K, Mauskop A (2004) Petasites hybridus root (butterbur) is an effective preventive treatment for migraine. Neurology 63:2240–2244

Lorenzetti BB, Ferreira SH (1985) Mode of analgesic action of dipyrone: direct antagonism of inflammatory hyperalgesia. Eur J Pharmacol 114:375–381

Lundberg JM, Franco-Cereceda A, Alving K, Delay-Goyet P, Lou YP (1992) Release of calcitonin gene-related peptide from sensory neurons. Ann N Y Acad Sci 657:187–193

Ma J, Altomare A, Rieder F, Behar J, Biancani P, Harnett KM (2011) ATP: a mediator for HCl-induced TRPV1 activation in esophageal mucosa. Am J Physiol Gastrointest Liver Physiol 301:G1075–G1082

Macpherson LJ, Dubin AE, Evans MJ, Marr F, Schultz PG, Cravatt BF, Patapoutian A (2007) Noxious compounds activate TRPA1 ion channels through covalent modification of cysteines. Nature 445:541–545

Maggi CA, Santicioli P, Geppetti P, Patacchini R, Frilli S, Astolfi M, Fusco B, Meli A (1988) Simultaneous release of substance P- and calcitonin gene-related peptide (CGRP)-like immunoreactivity from isolated muscle of the guinea pig urinary bladder. Neurosci Lett 87:163–167

Masterson CG, Durham PL (2010) DHE repression of ATP-mediated sensitization of trigeminal ganglion neurons. Headache 50:1424–1439

Materazzi S, Benemei S, Fusi C, Gualdani R, De Siena G, Vastani N, Andersson DA, Trevisan G, Moncelli MR, Wei X, Dussor G, Pollastro F, Patacchini R, Appendino G, Geppetti P, Nassini R (2013) Parthenolide inhibits nociception and neurogenic vasodilatation in the trigeminovascular system by targeting the TRPA1 channel. Pain 154:2750–2758

Mori Y, Takahashi N, Polat OK, Kurokawa T, Takeda N, Inoue M (2016) Redox-sensitive transient receptor potential channels in oxygen sensing and adaptation. Pflugers Arch 468:85–97

Moskowitz MA (1992) Neurogenic versus vascular mechanisms of sumatriptan and ergot alkaloids in migraine. Trends Pharmacol Sci 13:307–311

Mulderry PK, Ghatei MA, Spokes RA, Jones PM, Pierson AM, Hamid QA, Kanse S, Amara SG, Burrin JM, Legon S et al (1988) Differential expression of alpha-CGRP and beta-CGRP by primary sensory neurons and enteric autonomic neurons of the rat. Neuroscience 25:195–205

Nassini R, Materazzi S, Vriens J, Prenen J, Benemei S, De Siena G, la Marca G, Andre E, Preti D, Avonto C, Sadofsky L, Di Marzo V, De Petrocellis L, Dussor G, Porreca F, Taglialatela-Scafati O, Appendino G, Nilius B, Geppetti P (2012) The 'headache tree' via umbellulone and TRPA1 activates the trigeminovascular system. Brain 135:376–390

Nassini R, Fusi C, Materazzi S, Coppi E, Tuccinardi T, Marone IM, De Logu F, Preti D, Tonello R, Chiarugi A, Patacchini R, Geppetti P, Benemei S (2015) The TRPA1 channel mediates the analgesic action of dipyrone and pyrazolone derivatives. Br J Pharmacol 172:3397–3411

Nicoletti P, Trevisani M, Manconi M, Gatti R, De Siena G, Zagli G, Benemei S, Capone JA, Geppetti P, Pini LA (2008) Ethanol causes neurogenic vasodilation by TRPV1 activation and CGRP release in the trigeminovascular system of the Guinea pig. Cephalalgia 28:9–17

Nilius B, Szallasi A (2014) Transient receptor potential channels as drug targets: from the science of basic research to the art of medicine. Pharmacol Rev 66:676–814

Olesen J, Diener HC, Husstedt IW, Goadsby PJ, Hall D, Meier U, Pollentier S, Lesko LM (2004) Calcitonin gene-related peptide receptor antagonist BIBN 4096 BS for the acute treatment of migraine. N Engl J Med 350:1104–1110

Ooi L, Gigout S, Pettinger L, Gamper N (2013) Triple cysteine module within M-type K+ channels mediates reciprocal channel modulation by nitric oxide and reactive oxygen species. J Neurosci 33:6041–6046

Persson MG, Agvald P, Gustafsson LE (1994) Detection of nitric oxide in exhaled air during administration of nitroglycerin in vivo. Br J Pharmacol 111:825–828

Pothmann R, Danesch U (2005) Migraine prevention in children and adolescents: results of an open study with a special butterbur root extract. Headache 45:196–203

Pozsgai G, Hajna Z, Bagoly T, Boros M, Kemeny A, Materazzi S, Nassini R, Helyes Z, Szolcsanyi J, Pinter E (2012) The role of transient receptor potential ankyrin 1 (TRPA1) receptor activation in hydrogen-sulphide-induced CGRP-release and vasodilation. Eur J Pharmacol 689:56–64

Ramacciotti AS, Soares BG, Atallah AN (2007) Dipyrone for acute primary headaches. Cochrane Database Syst Rev 2:CD004842

Ramachandran R, Bhatt DK, Ploug KB, Hay-Schmidt A, Jansen-Olesen I, Gupta S, Olesen J (2014) Nitric oxide synthase, calcitonin gene-related peptide and NK-1 receptor mechanisms are involved in GTN-induced neuronal activation. Cephalalgia 34:136–147

Reuter U, Bolay H, Jansen-Olesen I, Chiarugi A, Sanchez del Rio M, Letourneau R, Theoharides TC, Waeber C, Moskowitz MA (2001) Delayed inflammation in rat meninges: implications for migraine pathophysiology. Brain 124:2490–2502

Rubio-Beltran E, Labastida-Ramirez A, Villalon CM, MaassenVanDenBrink A (2018) Is selective 5-HT1F receptor agonism an entity apart from that of the triptans in antimigraine therapy? Pharmacol Ther 17:30012–30013

Russell FA, King R, Smillie SJ, Kodji X, Brain SD (2014) Calcitonin gene-related peptide: physiology and pathophysiology. Physiol Rev 94:1099–1142

Sanada M, Yasuda H, Omatsu-Kanbe M, Sango K, Isono T, Matsuura H, Kikkawa R (2002) Increase in intracellular Ca2+ and calcitonin gene-related peptide release through metabotropic P2Y receptors in rat dorsal root ganglion neurons. Neuroscience 111:413–422

Sawada Y, Hosokawa H, Matsumura K, Kobayashi S (2008) Activation of transient receptor potential ankyrin 1 by hydrogen peroxide. Eur J Neurosci 27:1131–1142

Sicuteri F, Del Bene E, Poggioni M, Bonazzi A (1987) Unmasking latent dysnociception in healthy subjects. Headache 27:180–185

Silberstein SD, Dodick DW, Bigal ME, Yeung PP, Goadsby PJ, Blankenbiller T, Grozinski-Wolff M, Yang R, Ma Y, Aycardi E (2017) Fremanezumab for the preventive treatment of chronic migraine. N Engl J Med 377:2113–2122

Sinclair SR, Kane SA, Van der Schueren BJ, Xiao A, Willson KJ, Boyle J, de Lepeleire I, Xu Y, Hickey L, Denney WS, Li CC, Palcza J, Vanmolkot FH, Depre M, Van Hecken A, Murphy MG, Ho TW, de Hoon JN (2010) Inhibition of capsaicin-induced increase in dermal blood flow by the oral CGRP receptor antagonist, telcagepant (MK-0974). Br J Clin Pharmacol 69:15–22

Staikopoulos V, Sessle BJ, Furness JB, Jennings EA (2007) Localization of P2X2 and P2X3 receptors in rat trigeminal ganglion neurons. Neuroscience 144:208–216

Steinhoff M, Vergnolle N, Young SH, Tognetto M, Amadesi S, Ennes HS, Trevisani M, Hollenberg MD, Wallace JL, Caughey GH, Mitchell SE, Williams LM, Geppetti P, Mayer EA, Bunnett NW (2000) Agonists of proteinase-activated receptor 2 induce inflammation by a neurogenic mechanism. Nat Med 6:151–158

Steranka LR, Manning DC, DeHaas CJ, Ferkany JW, Borosky SA, Connor JR, Vavrek RJ, Stewart JM, Snyder SH (1988) Bradykinin as a pain mediator: receptors are localized to sensory neurons, and antagonists have analgesic actions. Proc Natl Acad Sci U S A 85:3245–3249

Strecker T, Dux M, Messlinger K (2002) Nitric oxide releases calcitonin-gene-related peptide from rat dura mater encephali promoting increases in meningeal blood flow. J Vasc Res 39:489–496

Tan LL, Bornstein JC, Anderson CR (2010) The neurochemistry and innervation patterns of extrinsic sensory and sympathetic nerves in the myenteric plexus of the C57Bl6 mouse jejunum. Neuroscience 166:564–579

Tfelt-Hansen P, Olesen J (2011) Possible site of action of CGRP antagonists in migraine. Cephalalgia 31:748–750

Thadani U, Rodgers T (2006) Side effects of using nitrates to treat angina. Expert Opin Drug Saf 5:667–674

Thomet OA, Wiesmann UN, Blaser K, Simon HU (2001) Differential inhibition of inflammatory effector functions by petasin, isopetasin and neopetasin in human eosinophils. Clin Exp Allergy 31:1310–1320

Tietjen GE, Herial NA, White L, Utley C, Kosmyna JM, Khuder SA (2009) Migraine and biomarkers of endothelial activation in young women. Stroke 40:2977–2982

Trainor DC, Jones RC (1966) Headaches in explosive magazine workers. Arch Environ Health Int J 12:231–234

Tramontana M, Giuliani S, Del Bianco E, Lecci A, Maggi CA, Evangelista S, Geppetti P (1993) Effects of capsaicin and 5-HT$_3$ antagonists on 5-hydroxytryptamine-evoked release of calcitonin gene-related peptide in the guinea-pig heart. Br J Pharmacol 108:431–435

Trevisani M, Smart D, Gunthorpe MJ, Tognetto M, Barbieri M, Campi B, Amadesi S, Gray J, Jerman JC, Brough SJ, Owen D, Smith GD, Randall AD, Harrison S, Bianchi A, Davis JB, Geppetti P (2002) Ethanol elicits and potentiates nociceptor responses via the vanilloid receptor-1. Nat Neurosci 5:546–551

Trevisani M, Siemens J, Materazzi S, Bautista DM, Nassini R, Campi B, Imamachi N, Andre E, Patacchini R, Cottrell GS, Gatti R, Basbaum AI, Bunnett NW, Julius D, Geppetti P (2007) 4-Hydroxynonenal, an endogenous aldehyde, causes pain and neurogenic inflammation through activation of the irritant receptor TRPA1. Proc Natl Acad Sci U S A 104:13519–13524

Vandewauw I, De Clercq K, Mulier M, Held K, Pinto S, Van Ranst N, Segal A, Voet T, Vennekens R, Zimmermann K, Vriens J, Voets T (2018) A TRP channel trio mediates acute noxious heat sensing. Nature 555:662–666

Zhang X, Kainz V, Zhao J, Strassman AM, Levy D (2013) Vascular extracellular signal-regulated kinase mediates migraine-related sensitization of meningeal nociceptors. Ann Neurol 73:741–750

Zhao Q, Liu Z, Wang Z, Yang C, Liu J, Lu J (2007) Effect of prepro-calcitonin gene-related peptide-expressing endothelial progenitor cells on pulmonary hypertension. Ann Thorac Surg 84:544–552

Zhu J, Miao XR, Tao KM, Zhu H, Liu ZY, Yu DW, Chen QB, Qiu HB, Lu ZJ (2017) Trypsin-protease activated receptor-2 signaling contributes to pancreatic cancer pain. Oncotarget 8:61810–61823

Zimmermann K, Reeh PW, Averbeck B (2002) ATP can enhance the proton-induced CGRP release through P2Y receptors and secondary PGE2 release in isolated rat dura mater. Pain 97:259–265

CGRP in Animal Models of Migraine

Anne-Sophie Wattiez, Mengya Wang, and Andrew F. Russo

Contents

A.-S. Wattiez
Department of Molecular Physiology and Biophysics, University of Iowa, Iowa City, IA, USA

Center for the Prevention and Treatment of Visual Loss, Iowa VA Health Care System, Iowa City, IA, USA

M. Wang
Department of Pharmacology, University of Iowa, Iowa City, IA, USA

A. F. Russo (✉)
Department of Molecular Physiology and Biophysics, University of Iowa, Iowa City, IA, USA

Center for the Prevention and Treatment of Visual Loss, Iowa VA Health Care System, Iowa City, IA, USA

Department of Neurology, University of Iowa, Iowa City, IA, USA
e-mail: andrew-russo@uiowa.edu

© Springer Nature Switzerland AG 2018
S. D. Brain, P. Geppetti (eds.), *Calcitonin Gene-Related Peptide (CGRP) Mechanisms*,
Handbook of Experimental Pharmacology 255, https://doi.org/10.1007/164_2018_187

Abstract

With the approval of calcitonin gene-related peptide (CGRP) and CGRP receptor monoclonal antibodies by the Federal Drug Administration, a new era in the treatment of migraine patients is beginning. However, there are still many unknowns in terms of CGRP mechanisms of action that need to be elucidated to allow new advances in migraine therapies. CGRP has been studied both clinically and preclinically since its discovery. Here we review some of the preclinical data regarding CGRP in animal models of migraine.

Keywords

Animal model · Antibody · CGRP · Migraine

1   Introduction

Migraine is the third most common medical condition in the world (Vos et al. 2012) and a highly debilitating neurological disease (Lipton et al. 2007). Calcitonin gene-related peptide (CGRP) has been in the forefront of migraine research for years, both clinically and in animal models. With the arrival of CGRP monoclonal antibodies for the treatment of migraine headaches, patients are hoping to find better relief for a disorder that highly impairs their quality of life. Here we will review some of the preclinical evidence that led to the realization that CGRP is a key player in migraine pathophysiology (Edvinsson et al. 2018; Ong et al. 2018; Russo 2015a).

After the initial discovery of CGRP (Amara et al. 1982; Rosenfeld et al. 1983), it was suggested that the neuropeptide was linked to nociception and cardiovascular regulation due to its distribution in small trigeminal and spinal sensory ganglion cells and in sensory fibers surrounding the blood vessels (Rosenfeld et al. 1983). Functional studies soon demonstrated that CGRP is the most potent vasodilatory peptide (Brain et al. 1985, 1986), a record that still stands today (Brain and Grant 2004; Russell et al. 2014). In the following years, more systematic studies of CGRP distribution (Skofitsch and Jacobowitz 1985), as well as its colocalization with substance P (Lee et al. 1985; Uddman et al. 1985; Wiesenfeld-Hallin et al. 1984), provided hints that CGRP might play a role in migraine pathophysiology (Edvinsson 1985). Studies in humans and animal models soon afterward laid the foundation for the field as we know it today (Edvinsson 2017).

2 Involvement of CGRP in Animal Migraine Models

Over the years there have been a variety of animal models developed to study migraine. Many, which are described below, were shown to involve CGRP and its receptor in some way. Since CGRP has a multitude of actions in the body (Russell et al. 2014), it is hard to predict which of these may be key for migraine. Nonetheless, evidence from animal models suggests there are both peripheral and central sensitization mechanisms that may be relevant to migraine (Russo 2015b). In the periphery, CGRP is released from primary afferents of the trigeminal nerve into the perivascular space of the meninges, as well as within the ganglia. Receptors have been identified on arterioles, primary afferents that do not express CGRP, glia, and mast cells (Fig. 1). Actions at some or all of these sites can lead to sensitization of trigeminal nociceptive fibers that could contribute to the headache of migraine. In the central nervous system, CGRP released from neurons can act as a neuromodulator to increase glutamatergic signaling. This enhanced neurotransmission could in turn lead to central sensitization that could contribute to headache and other heightened sensory perceptions, such as photophobia. These models are outlined in Fig. 1 and are described below.

2.1 Trigeminal Ganglion Activation Model

The subjective nature of headaches often precludes proper diagnosis and treatment but also makes this pathology hard to model in animals, which cannot orally report their pain. Based on the idea that migraine involves the activation of the trigeminovascular system, one of the first animal models for migraine headache was stimulation of the trigeminal ganglion. This model helped improve our understanding of the anatomy and pharmacology of the trigeminovascular system (Akerman et al. 2013). In particular, Goadsby et al. showed that the electrical stimulation of the trigeminal ganglion in cats induced an elevation of CGRP-like immunoreactivity in blood samples taken from the external jugular vein (Goadsby et al. 1988) and increased the release of CGRP into the cranial circulation on the side of the stimulation (Goadsby and Edvinsson 1993). In rats, stimulation of the trigeminal ganglion caused an increase in blood flow ipsilateral to the side of stimulation that was reduced by intravenous injection of the CGRP antagonist $CGRP_{8-37}$ (Escott et al. 1995).

2.2 Meningeal Stimulation Model

One of the most widely used models of migraine headache to date is stimulation of the dura mater of the meninges that line the brain. Meningeal stimulation can be achieved by application of inflammatory compounds or by electrical stimulation. In anesthetized cats, electrical stimulation of the superior sagittal sinus increased the levels of CGRP in jugular vein blood by 85%, which provided the first evidence that

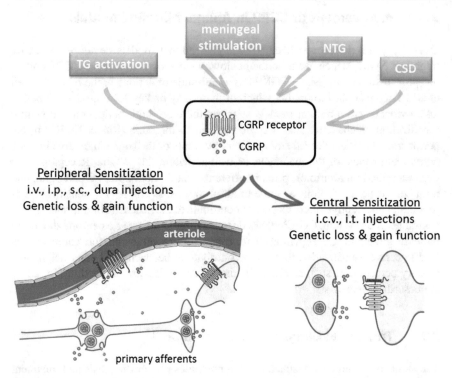

Fig. 1 CGRP in animal models of migraine. Animal models of migraine induced by activation of the trigeminal ganglia (TG), meningeal stimulation, infusion of nitroglycerin (NTG), or cortical spreading depression (CSD) have been shown to involve CGRP and its receptor. A schematic of the calcitonin-like receptor and RAMP1 complex is shown on vessels and neurons. Not shown are CGRP receptors on other cells, including mast cells and glia. While the exact sites and actions of CGRP that are important for migraine are not known, evidence from animal models suggests there are both peripheral and central sensitization mechanisms. Likewise, administration of CGRP by peripheral and central routes is believed to induce migraine-like phenotypes through these sensitization mechanisms. Peripheral delivery routes include intravenous (i.v.), intraperitoneal (i.p.), subcutaneous (s.c.), and directly onto the dura. Central delivery includes intracerebroventricular (i.c.v.) and intrathecal (i.t.) routes. Genetic models involving loss or gain of CGRP and/or receptor subunits can also modulate peripheral and central CGRP actions

activation of craniovascular afferents causes release of vasodilatory peptides (Zagami et al. 1990). Using immunohistochemistry, Messlinger and colleagues later showed that the parietal dura mater of the rat was densely innervated by CGRP nerve fibers (Messlinger et al. 1995). Furthermore, it was shown that electrical stimulation of the dural surface caused a depletion of CGRP-immunopositive fibers, suggesting a release of CGRP, and an increase of the dural blood flow around branches of the medial meningeal artery (Messlinger et al. 1995). It was concluded that the stimulation of trigeminal afferents innervating the dura mater induced the release of CGRP from peptidergic afferent terminals, which in turn caused vasodilation and increased meningeal blood flow.

This increase in meningeal blood flow was inhibited in a dose-dependent manner by topical application of the CGRP antagonist $CGRP_{8-37}$ (Kurosawa et al. 1995). In separate studies, Williamson et al. showed that $CGRP_{8-37}$ and two different $5HT_{1B/1D}$ agonists (sumatriptan and rizatriptan) were able to reduce the dilation of dural vessels induced by electrical stimulation in rats and guinea pigs (Williamson et al. 1997, 2001), which mimicked for the first time clinical findings that triptans were able to normalize CGRP levels during a migraine attack (Goadsby et al. 1990). Intravenous injection of another CGRP antagonist, BIBN4096BS, was able to prevent the vasodilatory actions of endogenous CGRP released following transcranial electrical stimulation in rats (Petersen et al. 2004; Troltzsch et al. 2007), as well as inhibit trigeminocervical superior sagittal sinus-evoked activity in cats (Storer et al. 2004). Taken together, those studies show that across species, there are CGRP receptors in the trigeminocervical complex. These data supported the hypothesis that blocking CGRP would be an effective treatment of migraine.

Using isolated rat middle cerebral arteries, CGRP was shown to induce a concentration-dependent dilation with abluminal application, but not by luminal application (Edvinsson et al. 2007). This suggested that CGRP could act on smooth muscle cell CGRP receptors, but could not cross the endothelial barrier. CGRP blockers such as $CGRP_{8-37}$, BIBN4096BS, and CGRP antibody were able to inhibit CGRP-induced relaxation (Edvinsson et al. 2007).

Meningeal stimulation can also be achieved by application of substances directly on the dura. A recognized symptom of migraine is heightened sensitivity to stimuli. The perception of touch as a painful stimulus is reported by nearly half of migraineurs (LoPinto et al. 2006; Mathew et al. 2004). In animal models, this mechanical allodynia can be measured using von Frey filaments. Application of inflammatory mediators directly onto the dura elicits both facial and plantar allodynia that can be reversed by sumatriptan and $CGRP_{8-37}$ (Edelmayer et al. 2009), once again showing that targeting CGRP is a valid strategy to treat pain associated with migraine.

More recently, a study investigated sex differences in behavioral responses after application of inflammatory soup (IS) on the dura (Stucky et al. 2011). While both male and female rats showed behavioral responses (activity measures as well as withdrawal responses for periorbital and perimasseter mechanical testing) to IS application compared to saline, females showed effects at lower doses than males and for longer duration. However, males required fewer applications of IS to exhibit responses (Stucky et al. 2011). In the same study, levels of transcripts for CGRP and the different subunits of its principal receptor (RAMP1, receptor activity-modifying protein 1; CLR, calcitonin-like receptor; and RCP, receptor component unit) were assessed in different areas of the CNS, at baseline or after application of IS to the dura. At baseline, females had lower levels of the receptor components in the trigeminal ganglion and in the medulla, while their CGRP mRNA levels were higher in the medulla than males. After IS and saline application to the dura, CGRP transcript levels were upregulated in all groups (Stucky et al. 2011). This suggests that the CGRP pathway responds to changes in intracranial pressure or meningeal stretch, while migraine-like behaviors occur after meningeal inflammation.

2.3 Nitroglycerin-Induced Model

Nitric oxide (NO) is a regulator of cerebral blood flow and vessel diameter. Nitroglycerin (NTG), a NO donor, can be used to trigger migraine in migraineurs (Christiansen et al. 1999; Thomsen et al. 1994), and response to nitroglycerin is a diagnostic test for migraine (Ferrari et al. 2015). This has led to the use of NTG administration as a trigger for sensory hypersensitivity associated with migraine in laboratory animals. Considering that NO is an important signaling molecule involved in the synthesis and release of CGRP from trigeminal ganglion neurons (Bowen et al. 2006), many of the drugs designed to block CGRP to treat migraine symptoms are also effective in NTG-induced migraine models.

NTG administration induces thermal and mechanical allodynia as well as thermal hyperalgesia in rodents (Bates et al. 2010; Tassorelli et al. 2003). Sumatriptan, the gold-standard anti-migraine drug is a 5-HT_{1B} and 5-HT_{1D} agonist that can prevent the release of CGRP in plasma (Goadsby and Edvinsson 1994). Administered centrally (i.t.) and peripherally (i.p.), sumatriptan was able to reduce NTG-induced thermal hypersensitivity; only the central injection of sumatriptan was able to reduce mechanical hypersensitivity (Bates et al. 2010). Similarly, NO-induced increase in spinal trigeminal activity can be reduced by the CGRP receptor antagonists BIBN4096BS (later called olcegepant) and MK-8825 (Feistel et al. 2013; Koulchitsky et al. 2009). Subsequently, it was found that nitroglycerin (i.p.) administration to rats increased CGRP levels in the brainstem and trigeminal ganglia (Capuano et al. 2014). Additionally, those authors showed that an injection of CGRP in the whisker pads of rats only increased the time the rats spent in face rubbing when they were pre-treated with NTG (Capuano et al. 2014). This suggests that NTG can sensitize the trigeminal system for CGRP to induce a painful behavior in rats (Capuano et al. 2014). In order to study the progression from acute to chronic migraine, Pradhan and colleagues used chronic injection of NTG every other day for 9 days, which induced progressive and sustained hyperalgesia (Pradhan et al. 2014). This time however, systemic or central sumatriptan did not ameliorate NTG-induced chronic hyperalgesia (Pradhan et al. 2014).

2.4 Cortical Spreading Depression Model

Cortical spreading depression (CSD) is hypothesized to cause migraine auras (Cutrer and Huerter 2007). There is some evidence pointing toward an interaction between CSD events and CGRP actions (Close et al. 2018). A recent study showed that BIBN4096BS could decrease the amplitude and propagation rate of repeated retinal spreading depression episodes induced by potassium in chick retinal preparation (Wang et al. 2016). Blocking CGRP receptors with $CGRP_{8-37}$ attenuated CSD-associated hyperperfusion in the rat (Reuter et al. 1998) and reduced CSD-induced pial dilatation (Colonna et al. 1994; Wahl et al. 1994), suggesting that the release of CGRP by trigeminal sensory neurons is responsible, at least in part, for some of the vascular changes associated with CSD. In a recent study,

MK-8825 (CGRP receptor antagonist) did not alter CSD waves or CSD-induced change in regional cerebral blood flow (Filiz et al. 2017). It did however attenuate CSD-induced trigeminal nerve-mediated freezing and spontaneous responses (both body and head grooming, wet dog shakes, and head shakes). Other behaviors such as eating/drinking, rearing, and turning that are impaired after induction of CSD were not changed after administration of MK-8825 (Filiz et al. 2017). Finally, CSD-induced periorbital allodynia is reversed by administration of MK-8825 (Filiz et al. 2017). Taken together, those studies show that the blockade of CGRP can decrease the impact of CSD and seem to indicate that CGRP may be involved in the propagation of CSD (but not its initiation). Interestingly, a recent study showed that induction of CSD by KCl in rats resulted in increased CGRP protein expression in the trigeminal ganglia, although there was no change in CGRP transcript levels (Yisarakun et al. 2015), which points to a possible positive feedback loop between CSD and CGRP (Close et al. 2018).

3 Animal Models of Migraine Induced by Injection of CGRP

After it was reported that (1) CGRP levels are elevated during spontaneous migraine and in between attacks in patients with chronic migraine (Bellamy et al. 2006; Goadsby et al. 1990; van Dongen et al. 2017), and (2) an intravenous infusion of CGRP could induce a delayed migraine-like headache in migraineurs (Asghar et al. 2011; Hansen et al. 2010; Lassen et al. 2002), animal models of migraine induced by injection of CGRP were developed. Administration of CGRP by peripheral and central routes is believed to induce migraine-like phenotypes through peripheral and central sensitization mechanisms, respectively (Fig. 1). However, it must be emphasized that the mechanisms are not mutually exclusive. For example, peripheral sensitization can lead to central sensitization. In this section, we describe animal models based on the peripheral and central delivery routes, which are outlined in Fig. 1 and summarized in Table 1.

3.1 Intravenous CGRP Delivery

Intravenous infusion of CGRP at a dose able to induce vasodilation is sufficient to induce a migraine-like headache in 66% of migraineurs (Asghar et al. 2011; Guo et al. 2016; Hansen et al. 2010; Lassen et al. 2002). In contrast, it only provokes a mild headache in non-migraineurs (Petersen et al. 2005a), suggesting that migraineurs are more sensitive to CGRP (Russo et al. 2009). Based on these clinical observations, the effects of i.v. CGRP have been studied in animals (Table 1).

Up until the discovery of one unique wild-type rat displaying spontaneous episodic trigeminal allodynia (Munro et al. 2018; Oshinsky et al. 2012), scientists had not been able to witness any occurrence of spontaneous migraine symptoms in laboratory animals. It was therefore not possible to discriminate a migraineur vs. non-migraineur population in animals without evoking symptoms. Nevertheless,

Table 1 Potential migraine-related effects of CGRP administered by different routes in rodents

CGRP delivery	Phenotype	Species and reference
Intravenous	• ↓ Blood pressure • No c-Fos activation in in trigeminal nucleus caudalis • ↑ c-Fos protein in the brainstem • ↑ p-ERK in dura mater	Rat Bhatt et al. (2015)
	• ↓ Mean arterial pressure • tachycardia • ↑ Cardiac output • No change in stroke volume • ↓ Total peripheral resistance	Rat Lappe et al. (1987)
	• ↓ Mean arterial pressure • ↑ Heart rate and cardiac output • ↓ Total peripheral resistance • ↑ Mesenteric and hindquarter blood flow • Dose-dependent changes in renal blood flow • ↓ Resistance in all vascular beds	Rat Siren and Feuerstein (1988)
	• Dilation of middle meningeal artery • Facilitated vibrissal responses	Rat Cumberbatch et al. (1999)
	• Hypotension • Dilation of middle meningeal artery • ↑ Pial artery/arteriole diameter	Rat Petersen et al. (2004)
	• ↑ Dilation of cerebral arteries when applied abluminally but not luminally • ↑ Dilation of cerebral cortical pial arteries/arterioles • ↓ Blood pressure	Rat Petersen et al. (2005a, b)
	• ↑ Dural blood flow • No activation or sensitization of meningeal nociceptors	Rat Levy et al. (2005)
Intracerebroventricular	• No c-Fos activation in trigeminal nucleus caudalis	Rat Bhatt et al. (2014)
	• ↑ Light aversion • ↓ Locomotion in the dark	Mouse Kaiser et al. (2012)
	• ↑ Light aversion • ↑ Resting in dark	Mouse Mason et al. (2017)
	• ↑ Hindpaw withdrawal latency to thermal and mechanical stimulation (antinociception)	Rat Huang et al. (2000)
	• ↑ Tail-flick latencies to thermal stimulation (antinociception) • ↑ Response latencies on the hot plate (antinociception) • ↓ Evoked thalamic neuronal firing	Rat Pecile et al. (1987)

(continued)

Table 1 (continued)

CGRP delivery	Phenotype	Species and reference
	• ↑ Paw-withdrawal latencies in C57BL/6 mice • ↓ Depression-like behavior in forced swim test in both C57BL/6 and AKR mice	Mouse Schorscher-Petcu et al. (2009)
Intrathecal	• ↑ Mechanical and thermal hyperalgesia in AKR but not C57BL/6 mice	Mouse Mogil et al. (2005)
	• ↑ Hyperalgesia to mechanical noxious stimuli (pinching the hind paw)	Rat Oku et al. (1987)
	• ↑ Mechanical allodynia at high dose	Mouse Marquez de Prado et al. (2009)
Intraperitoneal	• ↑ Light aversion in CD1 and C57BL/6J • ↑ Resting in dark in both strains	Mouse Mason et al. (2017)
	• ↑ Facial signs of discomfort in CD1 and C57BL/6J • ↑ Squint in both strains	Mouse Rea et al. (2018)
	• ↑ Diarrhea	Mouse Kaiser et al. (2017)
Dural and epidural	• ↑ Dural blood flow • No activation or sensitization of meningeal nociceptors	Rat Levy et al. (2005)
	• ↓ Climbing hutch and face grooming • ↑ Immobile behavior	Rat Yao et al. (2017)
	• ↑ Periorbital hypersensitivity in female mice	Mouse Burgos Vega et al. (2017)
Subcutaneous and intradermal	• ↑ Blood flow • No change in thermal hyperalgesia	Rat Chu et al. (2000)
	• No change in mechanical hyperalgesia	Rat Nakamura-Craig and Gill (1991)
	• ↓ Paw withdrawal threshold to noxious heat (hyperalgesia) • Strain-dependent hyperalgesia with the hot plate assay • ↑ Sensitivity to mechanical stimulation with von Frey filaments	Mouse Mogil et al. (2005)
	• No tactile allodynia	Mouse Marquez de Prado et al. (2009)

scientists have studied the effect of i.v. CGRP in preclinical settings. Considering that migraine has historically been considered a vascular disorder, vascular actions of CGRP are important to take into account. In animals, an infusion of CGRP induced a dose-dependent decrease in blood pressure and increase in heart rate (Bhatt et al. 2015; Lappe et al. 1987; Siren and Feuerstein 1988). IV CGRP administration in rats caused a dilation of the cortical pial arteries and arterioles and of the middle meningeal artery and increased local cortical cerebral blood flow, all of which could be inhibited by the CGRP receptor antagonist BIBN4096BS (Cumberbatch et al. 1999; Petersen et al. 2004, 2005b). Surprisingly, and to our knowledge, there is very little in the literature about nociceptive actions of i.v. CGRP. IV CGRP facilitated vibrissal responses, which seemed to indicate that CGRP-induced vasodilation was activating primary afferent meningeal nociceptors (Cumberbatch et al. 1999). However, electrophysiological studies later showed that it was not the case (Levy et al. 2005). Additionally, a recent study showed that CGRP infusion in awake rats failed to increase c-Fos and Zif268 (neuronal pain markers) expression in the trigeminal nucleus caudalis (Bhatt et al. 2015).

Although i.v. CGRP is the most translational approach for CGRP administration, the inherent difficulty and stress from performing an i.v. injection in rodents, especially in mice, led to the use of other routes of injections in preclinical studies.

3.2   Intraperitoneal CGRP Delivery

A relatively easy method to deliver CGRP to the peripheral tissues of rodents to allow assessment of migraine-like symptoms is by intraperitoneal (i.p.) injections (Table 1). Notably, i.p. CGRP induced light aversion both in CD1 and C57BL/6J mice, which was attenuated by both sumatriptan and an anti-CGRP antibody (Mason et al. 2017). These results, coupled with results obtained centrally, suggest that CGRP actions to induce migraine-like behavior are mediated by both peripheral and central mechanisms (Mason et al. 2017).

Recently, we also described i.p. CGRP-induced spontaneous pain in mice. The mice showed increased facial signs of discomfort (grimace and squint) (Rea et al. 2018). Those phenotypes were also reversed by anti-CGRP antibody. Interestingly, sumatriptan partially inhibited CGRP-induced spontaneous pain in males but not females (Rea et al. 2018). Of importance, the dose of 0.1 mg/kg i.p. used in all of our studies is able to induce vasodilation visible as redness of the ears (Rea et al. 2018).

Additionally, migraine symptomatology includes gastrointestinal problems, which occur in 22% of migraineurs (Kelman 2004). Our team reported that i.p. CGRP administration induced diarrhea in C57BL/6J mice and that olcegepant (previously called BIBN4096BS), a CGRP receptor antagonist, was able to attenuate this symptom (Kaiser et al. 2017).

The use of triptans in the previously mentioned studies validates the symptoms as being migraine-related but also provides some clues about triptan mechanisms of action. It is known that triptans are vasoconstrictors that can also inhibit endogenous neuropeptide release via 5-HT$_{1D}$ receptors (Durham and Russo 2002; Loder 2010).

Importantly, clinical studies demonstrated that triptans can reverse CGRP-induced vasodilation and headache in normal subjects and migraine patients (Asghar et al. 2010, 2011). Thus, in both mice and humans, triptans are able to override exogenous CGRP, suggesting that their mechanism of action must be more than just inhibition of CGRP release. Moreover, colocalization of 5-HT$_{1D}$ and CGRP in the spinal trigeminal nucleus and other areas in the brainstem such as the parabrachial nucleus (Noseda et al. 2008) and the fact that triptans can downregulate nociceptive signal transmission in the spinal trigeminal nucleus (Levy et al. 2005; Mitsikostas et al. 1999) support the hypothesis that triptans can mask the effect of a bolus injection of CGRP injected either centrally or peripherally. It is very likely that triptan mechanism of action to relieve migraine-like symptoms involves actions at multiple sites (Ahn and Basbaum 2005; Kaiser et al. 2012).

3.3 Subcutaneous and Intradermal CGRP Delivery

In accordance to the results obtained with other routes of administration, intradermal CGRP (in rats and in rabbits) induced an increase in blood flow (Brain et al. 1985; Chu et al. 2000) (Table 1). However, intradermal CGRP did not induce any thermal hyperalgesia in rats (Chu et al. 2000). Early studies also showed a lack of effect of subplantar CGRP compared to that of substance P and neurokinin A in exacerbating the response to paw pressure (mechanical hyperalgesia) in Wistar rats (Nakamura-Craig and Gill 1991). This is consistent with early results in the human skin where CGRP was proposed to have a role in blood flow regulation and in mediating flare response but most likely had no direct role in nociception since the concentrations at which it induced histamine release exceeded normal physiologic concentrations (Brain et al. 1986; Weidner et al. 2000). In a later study, Mogil and colleagues reported a strain difference in the development of thermal hyperalgesia after subcutaneous (s.c.) CGRP injection into the plantar hindpaw. AKR mice but not in C57BL/6J mice seemed to become hypersensitive (Mogil et al. 2005). In our hands, C57BL/6J mice did not develop tactile allodynia assessed by von Frey filaments after intraplantar injection of CGRP (Marquez de Prado et al. 2009). Since we have shown that CD1 mice are more sensitive to CGRP-induced light aversion (Mason et al. 2017), this lack of effect may be strain specific.

3.4 Dural and Epidural CGRP Delivery

Similar to previously described models of meningeal stimulation, and because the activation of the trigeminal nerve leads to release of CGRP from perivascular nerve endings at meningeal blood vessels, dural/epidural delivery of CGRP was used as a model for migraine pathophysiology (Table 1). Dural delivery of CGRP induced a significant increase in dural blood flow, although it reportedly did not activate or sensitize meningeal nociceptors (Levy et al. 2005). Recently however, Yao and colleagues described a reduction in climbing and face-grooming behaviors accompanied by increased immobile behavior after epidural CGRP administration in rats (Yao et al. 2017). Very interestingly, Dussor and colleagues have recently reported that CGRP can directly stimulate the dura of female but not male rodents to induce periorbital hypersensitivity (Burgos Vega et al. 2017). Additionally, CGRP was able to prime female rodents to a usually innocuous dural application of a pH 7.0 solution.

3.5 Intracerebroventricular CGRP Delivery

Since studies pointed towards a central mechanism of CGRP in migraine pathophysiology, intracerebroventricular (i.c.v.) injections of CGRP were studied (Table 1). Although i.c.v. injection in awake rats did not increase c-Fos expression in the trigeminal nucleus caudalis (Bhatt et al. 2014), a similar injection showed migraine-like behavioral effects. An important trigger and/or symptom experienced by migraineurs is photophobia or photosensitivity, which is an altered perception of light that elicits discomfort (Boulloche et al. 2010; Kelman 2007; Martin and Behbehani 2001; Mulleners et al. 2001; Rasmussen 1993; Spierings et al. 2001). In rodents, light aversion represents a surrogate for photophobia and can be measured using a conflict assay between a light and a dark chamber. Using this test, our lab has shown that i.c.v. injection of CGRP in mice induced light-aversive behavior when exposed to bright light (27,000 lux) but not to dim light (55 lux), which was attenuated by rizatriptan, a 5-HT$_{1B/D}$ agonist anti-migraine drug (Kaiser et al. 2012; Mason et al. 2017). Those animals also showed an increased time spent resting in the dark compartment, which is a behavior similar to migraineurs who tend to seek a dark place to rest during attacks (Kaiser et al. 2012; Mason et al. 2017). These findings indicate that CGRP can act in the CNS to cause light aversion.

Other studies assessed the role of i.c.v. CGRP on nociception. Antinociceptive effects of CGRP administered intracerebroventricularly into the nucleus raphe magnus, amygdala, nucleus accumbens, or the periaqueductal gray were reported in rats submitted to thermal and mechanical stimulations (Huang et al. 2000; Li et al. 2001; Pecile et al. 1987; Xu et al. 2003; Yu et al. 2003; Zhou et al. 2003). Additionally, it was reported that i.c.v. CGRP increased paw withdrawal latencies to thermal stimuli in C57BL/6 mice but not in AKR mice while decreasing depression-like behaviors in both strains in the forced swim test (Schorscher-Petcu et al. 2009). In the same study, i.c.v. CGRP and CGRP receptor antagonists failed to modulate activity in the elevated plus maze, a model of anxiety (Schorscher-Petcu et al. 2009).

3.6 Intrathecal CGRP Delivery

Studies showed that CGRP is located in small diameter dorsal root ganglion neurons (Hokfelt et al. 1992), dorsal horns (Hokfelt et al. 1992; Ishida-Yamamoto and Tohyama 1989), and intermediolateral and ventral horns of the spinal cord (Bennett et al. 2000; Marti et al. 1987; Senba and Tohyama 1988). Spinal cord central sensitization following intradermal capsaicin injection has been shown to be mediated by CGRP and its receptors (Carlton et al. 1990; Sun et al. 2004). Thus, researchers explored the effects of intrathecal (i.t.) CGRP administration on pain responses (Table 1). Administration of i.t. CGRP induced mechanical and thermal hyperalgesia in rats (Mogil et al. 2005; Oku et al. 1987). Our team showed that a low dose of CGRP injected into the lumbar spinal region did not exacerbate the response to an innocuous mechanical stimulus (von Frey filaments), while a higher dose evoked mechanical allodynia (Marquez de Prado et al. 2009). These data indicate that CGRP can act centrally to sensitize mice to thermal and mechanical stimuli.

4 Genetic Manipulation of CGRP in Migraine Models

Genetic manipulations have been used to directly investigate CGRP signaling. Transgenic and knockout mice have been generated that have either a loss or gain of function of the CGRP ligand or receptor subunits. These genetic models allow modulation of peripheral and central CGRP actions (Fig. 1). A thorough review of all CGRP and receptor subunit mutant mice and their phenotypes can be found elsewhere (Sowers et al. 2017). We will focus on the few cases where migraine-like phenotypes were assessed.

4.1 Overexpression of Human RAMP1

With the goal to study migraine, our laboratory developed a subset of genetic constructs revolving around the overexpression of the human receptor activity-modifying protein 1 (RAMP1) subunit of the CGRP receptor. Using RAMP1, which has been shown to be functionally rate-limiting (Zhang et al. 2006, 2007), allowed the development of CGRP-sensitized mice. Overexpressing the human version of the gene provided the advantage that it can be targeted by drugs designed for clinical application and therefore allow more translatable models (Russo 2015b). To generate those models, the approach was to use double-transgenic mice that express hRAMP1 in a tissue-specific Cre-dependent manner or in all tissues.

Global overexpression of hRAMP1 in all tissues was achieved using mice expressing Cre recombinase under the control of the ubiquitous adenovirus EIIa promoter (Bohn et al. 2017). Cultures obtained from vascular smooth muscle and trigeminal ganglia from those global mice showed an increased CGRP receptor activity that could be blocked by drugs such as CGRP receptor antagonists telcagepant and $CGRP_{8-37}$ (Bohn et al. 2017). Mice with global hRAMP1 overexpression display increased vasodilation of the carotid and basilar arteries, and cerebral arterioles after CGRP application (Chrissobolis et al. 2010), and decreased angiotensin II-induced hypertension (Sabharwal et al. 2010).

In order to study more specifically the role of CGRP and its receptor in the nervous system, double-transgenic mice were developed using nestin-Cre to drive expression in neurons and some glia cells (Zhang et al. 2007). The injection of CGRP in the whisker pads of those animals increased neurogenic inflammation by doubling plasma extravasation (Zhang et al. 2007). Those animals also display behaviors consistent with migraine such as mechanical allodynia and photosensitivity. While the *nestin/hRAMP1* mice have similar hindpaw withdrawal thresholds to von Frey filament stimulation than control littermates, their response frequency drastically increased after intrathecal CGRP injection, while the same dose of CGRP did not elicit a response in control animals (Marquez de Prado et al. 2009). *Nestin/hRAMP1* mice also show an increased sensitivity to tactile stimulation after capsaicin injection which extended to the contralateral hindpaw, suggesting central sensitization (Marquez de Prado et al. 2009). The transgenic *nestin/hRAMP1* mice display light-aversive behavior when confronted to bright light (Recober et al. 2009). This light aversion is enhanced after i.c.v. injection of CGRP even when exposed to very dim light (55 lux) (Recober et al. 2010). In the same conditions, those mice also display a decrease in motility behaviors once in the dark, such as rearing, distance travelled, time spent moving, and ambulatory velocity (Recober et al. 2010), which resembles the behavior of migraineurs who will seek out a dark room to rest during an attack.

As mentioned earlier, i.p. injection of CGRP in wild-type mice induced light aversion when exposed to very bright light (Mason et al. 2017). Interestingly, and contrasting to the results obtained with i.c.v. CGRP, *nestin/hRAMP1* transgenic mice were not sensitized to i.p. CGRP when exposed to dim lights (Mason et al. 2017). In conclusion, the hRAMP1 double transgenic mice enabled the understanding that CGRP is a key player in migraine both centrally through action on neurons and peripherally on receptors that are not located in the nervous system. Experiments are currently underway to assess the role of CGRP receptors on smooth muscle and the endothelium in the periphery.

4.2   Other Transgenic Models

A few other transgenic models affecting CGRP signaling assessed nociceptive and vascular changes that can have implications for migraine pathophysiology.

In terms of nociception, different lines of CGRP knockout mice have been developed that show maladaptation to pain. In contrast to wild-type mice, Zhang and colleagues reported a CT/αCGRP knockout mouse that showed no sign of secondary hyperalgesia after development of carrageenan-induced inflammation in the knee joint (Zhang et al. 2001). Another strain of αCGRP knockout showed an attenuated licking response to capsaicin and formalin injections as well as a reduction of the edema produced by carrageenan injection in the hindpaw (Salmon et al. 2001). This transgenic mouse also displayed no sign of thermal hyperalgesia after ATP-induced TRPV1 potentiation (Devesa et al. 2014) and reduced morphine analgesia (Salmon et al. 1999). CGRP knockout mice also present a reduced vestibule-ocular reflex (Luebke et al. 2014) and abnormal cochlear response (Maison et al. 2003) which can be of importance in the pathophysiology of migraine. Keeping in mind that migraine has a vascular component, the effect of CGRP gene deletion on the cardiovascular system was assessed but remains controversial, with reports of a lack of effect (Lu et al. 1999) and reports of increased blood pressure (Gangula et al. 2000; Oh-hashi et al. 2001). In one study, RAMP1 knockout mice also had elevated blood pressure (Tsujikawa et al. 2007).

5 CGRP Antibodies: New Era in Migraine Treatment

Monoclonal antibodies that target either CGRP or its receptor have now been approved by the Federal Drug Administration for the preventive treatment of migraine. Erenumab (Amgen/Novartis) blocks CGRP receptors. Fremanezumab (Teva Pharmaceuticals) and galcanezumab (Eli Lilly) bind to CGRP and block its binding to the receptors. A fourth antibody, eptinezumab (Alder Biopharmaceuticals), also blocks CGRP and is on track for approval.

In the 1980s and 1990s, it was found that intrathecal injection of CGRP antisera could block the pain induced by thermal (Kawamura et al. 1989) and mechanical (Kawamura et al. 1989; Kuraishi et al. 1988) noxious stimuli in rats receiving injections of adjuvant arthritis or carrageenin in the paw. In addition, CGRP antiserum partially rescued the reduced nociceptive threshold evoked by repeated cold stress (Satoh et al. 1992). However, antibody studies that are more relevant to migraine have only been pursued in the past few years.

Several studies have examined the effect of CGRP-blocking antibodies on migraine-like symptoms in mice. Mason et al. studied the effect of one monoclonal anti-CGRP antibody (ALD405) in light aversion in mice (Mason et al. 2017). These mice were first treated with CGRP (i.p.) to establish the degree of their responsiveness to CGRP. The mice were then given anti-CGRP antibody (i.p.) 24 h before given CGRP (i.p.) a second time. The amount of anti-CGRP antibody injected was ~eightfold excess over exogenous CGRP. The results showed that CGRP antibody attenuated CGRP (i.p.)-induced light-aversive behavior (details about CGRP-induced light aversion in Sect. 3.2). The results suggest a peripheral action of CGRP in the induction of light aversion. Likewise, Rea et al. showed that ALD405 administration (i.p.) prevented CGRP (i.p.)-induced spontaneous grimace (indicator of facial discomfort) in CD1 mice both in males and females. ALD405 administration also prevented the grimace in restrained C57BL/6J mice independent of the light. Another measurement of facial discomfort is squint, which was the principle component of the grimace, accounting for 77% of the total variation of grimace scale. CGRP-induced squint in restrained CD1 mice and C57BL/6J mice was prevented by ALD405 administration (i.p.). This suggests that CGRP can act in the periphery to induce a pain response. Gastrointestinal issues are one of the most common symptoms of migraine including diarrhea (Kelman 2004) (see Sect. 3.4). CGRP injection induced diarrhea in C57BL/6J mice, while anti-CGRP antibodies blocked CGRP-induced diarrhea (Kaiser et al. 2017). Moreover, it has been reported that a different CGRP antibody suppressed CSD as indicated by increased latency of CSD, and this effect was blocked by exogenous CGRP (Jiang et al. 2018).

In a series of experiments in rats, i.v. administration of the CGRP-blocking antibody fremanezumab was shown to inhibit the activation of high-threshold trigeminovascular neurons that were responsive to mechanical stimulation of the dura, but not to either innocuous or noxious stimulation of the skin or cornea. Fremanezumab also prevented the activation of trigeminovascular high-threshold neurons by CSD induced mechanically by inserting a glass micropipette into the visual cortex (Melo-Carrillo et al. 2017a). Moreover, fremanezumab pretreatment inhibited the response of Aδ, but not C-fiber, neurons in response to CSD (Melo-Carrillo et al. 2017b). These results provide a mechanism by which fremanezumab could reduce the intracranial pain of migraine. In addition, it was demonstrated that fremanezumab can treat medication overuse headache symptoms in rats. For these experiments, rats were primed with repeated sumatriptan or morphine treatments. Fremanezumab significantly inhibited bright-light stress or NO donor-induced cutaneous allodynia (Kopruszinski et al. 2017). The data suggest that medication overuse headache may be CGRP-dependent and that the anti-CGRP antibody may be a potential therapeutic.

6 Conclusion

In conclusion, many animal models of migraine involve CGRP to some degree. The importance of CGRP in those models is confirmed by the ability of direct injection of CGRP to induce several migraine-like symptoms in rodents. Further, these preclinical observations are in full alignment with the recent success of CGRP-based migraine therapeutics in patients. Thus, CGRP in animal models of migraine is an excellent example of successful translation of science from the lab to the patient.

Acknowledgments The authors thank members of the lab for their comments and support from Dept. Defense W81XWH-16-1-0071, W81XWH-16-1-0211; VA-ORD (RR&D) 1IO1RX002101; C6810-C and NIH NS075599.

References

Ahn AH, Basbaum AI (2005) Where do triptans act in the treatment of migraine? Pain 115:1–4. https://doi.org/10.1016/j.pain.2005.03.008

Akerman S, Holland PR, Hoffmann J (2013) Pearls and pitfalls in experimental in vivo models of migraine: dural trigeminovascular nociception. Cephalalgia 33:577–592. https://doi.org/10.1177/0333102412472071

Amara SG, Jonas V, Rosenfeld MG, Ong ES, Evans RM (1982) Alternative RNA processing in calcitonin gene expression generates mRNAs encoding different polypeptide products. Nature 298:240–244

Asghar MS et al (2010) Dilation by CGRP of middle meningeal artery and reversal by sumatriptan in normal volunteers. Neurology 75:1520–1526. https://doi.org/10.1212/WNL.0b013e3181f9626a

Asghar MS et al (2011) Evidence for a vascular factor in migraine. Ann Neurol 69:635–645. https://doi.org/10.1002/ana.22292

Bates EA et al (2010) Sumatriptan alleviates nitroglycerin-induced mechanical and thermal allodynia in mice. Cephalalgia 30:170–178. https://doi.org/10.1111/j.1468-2982.2009.01864.x

Bellamy JL, Cady RK, Durham PL (2006) Salivary levels of CGRP and VIP in rhinosinusitis and migraine patients. Headache 46:24–33. https://doi.org/10.1111/j.1526-4610.2006.00294.x

Bennett AD, Chastain KM, Hulsebosch CE (2000) Alleviation of mechanical and thermal allodynia by CGRP(8-37) in a rodent model of chronic central pain. Pain 86:163–175

Bhatt DK, Gupta S, Ploug KB, Jansen-Olesen I, Olesen J (2014) mRNA distribution of CGRP and its receptor components in the trigeminovascular system and other pain related structures in rat brain, and effect of intracerebroventricular administration of CGRP on Fos expression in the TNC. Neurosci Lett 559:99–104. https://doi.org/10.1016/j.neulet.2013.11.057

Bhatt DK, Ramachandran R, Christensen SL, Gupta S, Jansen-Olesen I, Olesen J (2015) CGRP infusion in unanesthetized rats increases expression of c-Fos in the nucleus tractus solitarius and caudal ventrolateral medulla, but not in the trigeminal nucleus caudalis. Cephalalgia 35:220–233. https://doi.org/10.1177/0333102414535995

Bohn KJ et al (2017) CGRP receptor activity in mice with global expression of human receptor activity modifying protein 1. Br J Pharmacol 174:1826–1840. https://doi.org/10.1111/bph.13783

Boulloche N, Denuelle M, Payoux P, Fabre N, Trotter Y, Geraud G (2010) Photophobia in migraine: an interictal PET study of cortical hyperexcitability and its modulation by pain. J Neurol Neurosurg Psychiatry 81:978–984. https://doi.org/10.1136/jnnp.2009.190223

Bowen EJ, Schmidt TW, Firm CS, Russo AF, Durham PL (2006) Tumor necrosis factor-alpha stimulation of calcitonin gene-related peptide expression and secretion from rat trigeminal ganglion neurons. J Neurochem 96:65–77. https://doi.org/10.1111/j.1471-4159.2005.03524.x

Brain SD, Grant AD (2004) Vascular actions of calcitonin gene-related peptide and adrenomedullin. Physiol Rev 84:903–934. https://doi.org/10.1152/physrev.00037.2003

Brain SD, Williams TJ, Tippins JR, Morris HR, MacIntyre I (1985) Calcitonin gene-related peptide is a potent vasodilator. Nature 313:54–56

Brain SD, Tippins JR, Morris HR, MacIntyre I, Williams TJ (1986) Potent vasodilator activity of calcitonin gene-related peptide in human skin. J Invest Dermatol 87:533–536

Burgos Vega C, Quigley L, Patel M, Price T, Arkopian A, Dussor G (2017) Meningeal application of prolactin and CGRP produces female specific migraine-related behavior in rodents. J Pain 18:S11

Capuano A, Greco MC, Navarra P, Tringali G (2014) Correlation between algogenic effects of calcitonin-gene-related peptide (CGRP) and activation of trigeminal vascular system, in an in vivo experimental model of nitroglycerin-induced sensitization. Eur J Pharmacol 740:97–102. https://doi.org/10.1016/j.ejphar.2014.06.046

Carlton SM, Westlund KN, Zhang DX, Sorkin LS, Willis WD (1990) Calcitonin gene-related peptide containing primary afferent fibers synapse on primate spinothalamic tract cells. Neurosci Lett 109:76–81

Chrissobolis S, Zhang Z, Kinzenbaw DA, Lynch CM, Russo AF, Faraci FM (2010) Receptor activity-modifying protein-1 augments cerebrovascular responses to calcitonin gene-related peptide and inhibits angiotensin II-induced vascular dysfunction. Stroke 41:2329–2334. https://doi.org/10.1161/STROKEAHA.110.589648

Christiansen I, Thomsen LL, Daugaard D, Ulrich V, Olesen J (1999) Glyceryl trinitrate induces attacks of migraine without aura in sufferers of migraine with aura. Cephalalgia 19:660–667. Discussion 626. https://doi.org/10.1046/j.1468-2982.1999.019007660.x

Chu DQ, Choy M, Foster P, Cao T, Brain SD (2000) A comparative study of the ability of calcitonin gene-related peptide and adrenomedullin(13-52) to modulate microvascular but not thermal hyperalgesia responses. Br J Pharmacol 130:1589–1596. https://doi.org/10.1038/sj.bjp.0703502

Close LN, Eftekhari S, Wang M, Charles AC, Russo AF (2018) Cortical spreading depression as a site of origin for migraine: role of CGRP. Cephalalgia 333102418774299. https://doi.org/10.1177/0333102418774299

Colonna DM, Meng W, Deal DD, Busija DW (1994) Calcitonin gene-related peptide promotes cerebrovascular dilation during cortical spreading depression in rabbits. Am J Physiol 266: H1095–H1102. https://doi.org/10.1152/ajpheart.1994.266.3.H1095

Cumberbatch MJ, Williamson DJ, Mason GS, Hill RG, Hargreaves RJ (1999) Dural vasodilation causes a sensitization of rat caudal trigeminal neurones in vivo that is blocked by a 5-HT1B/1D agonist. Br J Pharmacol 126:1478–1486. https://doi.org/10.1038/sj.bjp.0702444

Cutrer FM, Huerter K (2007) Migraine aura. Neurologist 13:118–125. https://doi.org/10.1097/01. nrl.0000252943.82792.38

Devesa I, Ferrandiz-Huertas C, Mathivanan S, Wolf C, Lujan R, Changeux JP, Ferrer-Montiel A (2014) alphaCGRP is essential for algesic exocytotic mobilization of TRPV1 channels in peptidergic nociceptors. Proc Natl Acad Sci U S A 111:18345–18350. https://doi.org/10. 1073/pnas.1420252111

Durham PL, Russo AF (2002) New insights into the molecular actions of serotonergic antimigraine drugs. Pharmacol Ther 94:77–92

Edelmayer RM et al (2009) Medullary pain facilitating neurons mediate allodynia in headache-related pain. Ann Neurol 65:184–193. https://doi.org/10.1002/ana.21537

Edvinsson L (1985) Functional role of perivascular peptides in the control of cerebral circulation. Trends Neurosci 8:126–131

Edvinsson L (2017) The trigeminovascular pathway: role of CGRP and CGRP receptors in migraine. Headache 57(Suppl 2):47–55. https://doi.org/10.1111/head.13081

Edvinsson L, Nilsson E, Jansen-Olesen I (2007) Inhibitory effect of BIBN4096BS, CGRP(8-37), a CGRP antibody and an RNA-Spiegelmer on CGRP induced vasodilatation in the perfused and non-perfused rat middle cerebral artery. Br J Pharmacol 150:633–640. https://doi.org/10.1038/ sj.bjp.0707134

Edvinsson L, Haanes KA, Warfvinge K, Krause DN (2018) CGRP as the target of new migraine therapies – successful translation from bench to clinic. Nat Rev Neurol 14:338–350. https://doi. org/10.1038/s41582-018-0003-1

Escott KJ, Beattie DT, Connor HE, Brain SD (1995) Trigeminal ganglion stimulation increases facial skin blood flow in the rat: a major role for calcitonin gene-related peptide. Brain Res 669:93–99

Feistel S, Albrecht S, Messlinger K (2013) The calcitonin gene-related peptide receptor antagonist MK-8825 decreases spinal trigeminal activity during nitroglycerin infusion. J Headache Pain 14:93. https://doi.org/10.1186/1129-2377-14-93

Ferrari MD, Klever RR, Terwindt GM, Ayata C, van den Maagdenberg AM (2015) Migraine pathophysiology: lessons from mouse models and human genetics. Lancet Neurol 14:65–80. https://doi.org/10.1016/S1474-4422(14)70220-0

Filiz A, Tepe N, Eftekhari S, Boran HE, Dilekoz E, Edvinsson L, Bolay H (2017) CGRP receptor antagonist MK-8825 attenuates cortical spreading depression induced pain behavior. Cephalalgia 333102417735845. https://doi.org/10.1177/0333102417735845

Gangula PR et al (2000) Increased blood pressure in alpha-calcitonin gene-related peptide/calcitonin gene knockout mice. Hypertension 35:470–475

Goadsby PJ, Edvinsson L (1993) The trigeminovascular system and migraine: studies characterizing cerebrovascular and neuropeptide changes seen in humans and cats. Ann Neurol 33:48–56. https://doi.org/10.1002/ana.410330109

Goadsby PJ, Edvinsson L (1994) Human in vivo evidence for trigeminovascular activation in cluster headache. Neuropeptide changes and effects of acute attacks therapies. Brain 117 (Pt 3):427–434

Goadsby PJ, Edvinsson L, Ekman R (1988) Release of vasoactive peptides in the extracerebral circulation of humans and the cat during activation of the trigeminovascular system. Ann Neurol 23:193–196. https://doi.org/10.1002/ana.410230214

Goadsby PJ, Edvinsson L, Ekman R (1990) Vasoactive peptide release in the extracerebral circulation of humans during migraine headache. Ann Neurol 28:183–187. https://doi.org/10. 1002/ana.410280213

Guo S, Vollesen ALH, Olesen J, Ashina M (2016) Premonitory and non-headache symptoms induced by CGRP and PACAP38 in patients with migraine. Pain 157:2773–2781. https://doi.org/10.1097/j.pain.0000000000000702

Hansen JM, Hauge AW, Olesen J, Ashina M (2010) Calcitonin gene-related peptide triggers migraine-like attacks in patients with migraine with aura. Cephalalgia 30:1179–1186. https://doi.org/10.1177/0333102410368444

Hokfelt T et al (1992) Calcitonin gene-related peptide in the brain, spinal cord, and some peripheral systems. Ann N Y Acad Sci 657:119–134

Huang Y, Brodda-Jansen G, Lundeberg T, Yu LC (2000) Anti-nociceptive effects of calcitonin gene-related peptide in nucleus raphe magnus of rats: an effect attenuated by naloxone. Brain Res 873:54–59

Ishida-Yamamoto A, Tohyama M (1989) Calcitonin gene-related peptide in the nervous tissue. Prog Neurobiol 33:335–386

Jiang L, Wang Y, Xu Y, Ma D, Wang M (2018) The transient receptor potential ankyrin type 1 plays a critical role in cortical spreading depression. Neuroscience 382:23–34. https://doi.org/10.1016/j.neuroscience.2018.04.025

Kaiser EA, Kuburas A, Recober A, Russo AF (2012) Modulation of CGRP-induced light aversion in wild-type mice by a 5-HT(1B/D) agonist. J Neurosci 32:15439–15449. https://doi.org/10.1523/JNEUROSCI.3265-12.2012

Kaiser EA, Rea BJ, Kuburas A, Kovacevich BR, Garcia-Martinez LF, Recober A, Russo AF (2017) Anti-CGRP antibodies block CGRP-induced diarrhea in mice. Neuropeptides 64:95–99. https://doi.org/10.1016/j.npep.2016.11.004

Kawamura M, Kuraishi Y, Minami M, Satoh M (1989) Antinociceptive effect of intrathecally administered antiserum against calcitonin gene-related peptide on thermal and mechanical noxious stimuli in experimental hyperalgesic rats. Brain Res 497:199–203

Kelman L (2004) The premonitory symptoms (prodrome): a tertiary care study of 893 migraineurs. Headache 44:865–872. https://doi.org/10.1111/j.1526-4610.2004.04168.x

Kelman L (2007) The triggers or precipitants of the acute migraine attack. Cephalalgia 27:394–402. https://doi.org/10.1111/j.1468-2982.2007.01303.x

Kopruszinski CM et al (2017) Prevention of stress- or nitric oxide donor-induced medication overuse headache by a calcitonin gene-related peptide antibody in rodents. Cephalalgia 37:560–570. https://doi.org/10.1177/0333102416650702

Koulchitsky S, Fischer MJ, Messlinger K (2009) Calcitonin gene-related peptide receptor inhibition reduces neuronal activity induced by prolonged increase in nitric oxide in the rat spinal trigeminal nucleus. Cephalalgia 29:408–417. https://doi.org/10.1111/j.1468-2982.2008.01745.x

Kuraishi Y, Nanayama T, Ohno H, Minami M, Satoh M (1988) Antinociception induced in rats by intrathecal administration of antiserum against calcitonin gene-related peptide. Neurosci Lett 92:325–329

Kurosawa M, Messlinger K, Pawlak M, Schmidt RF (1995) Increase of meningeal blood flow after electrical stimulation of rat dura mater encephali: mediation by calcitonin gene-related peptide. Br J Pharmacol 114:1397–1402

Lappe RW, Slivjak MJ, Todt JA, Wendt RL (1987) Hemodynamic effects of calcitonin gene-related peptide in conscious rats. Regul Pept 19:307–312

Lassen LH, Haderslev PA, Jacobsen VB, Iversen HK, Sperling B, Olesen J (2002) CGRP may play a causative role in migraine. Cephalalgia 22:54–61. https://doi.org/10.1046/j.1468-2982.2002.00310.x

Lee Y et al (1985) Coexistence of calcitonin gene-related peptide and substance P-like peptide in single cells of the trigeminal ganglion of the rat: immunohistochemical analysis. Brain Res 330:194–196

Levy D, Burstein R, Strassman AM (2005) Calcitonin gene-related peptide does not excite or sensitize meningeal nociceptors: implications for the pathophysiology of migraine. Ann Neurol 58:698–705. https://doi.org/10.1002/ana.20619

Li N, Lundeberg T, Yu LC (2001) Involvement of CGRP and CGRP1 receptor in nociception in the nucleus accumbens of rats. Brain Res 901:161–166

Lipton RB, Bigal ME, Diamond M, Freitag F, Reed ML, Stewart WF, Group AA (2007) Migraine prevalence, disease burden, and the need for preventive therapy. Neurology 68:343–349. https://doi.org/10.1212/01.wnl.0000252808.97649.21

Loder E (2010) Triptan therapy in migraine. N Engl J Med 363:63–70. https://doi.org/10.1056/NEJMct0910887

LoPinto C, Young WB, Ashkenazi A (2006) Comparison of dynamic (brush) and static (pressure) mechanical allodynia in migraine. Cephalalgia 26:852–856. https://doi.org/10.1111/j.1468-2982.2006.01121.x

Lu JT et al (1999) Mice lacking alpha-calcitonin gene-related peptide exhibit normal cardiovascular regulation and neuromuscular development. Mol Cell Neurosci 14:99–120. https://doi.org/10.1006/mcne.1999.0767

Luebke AE, Holt JC, Jordan PM, Wong YS, Caldwell JS, Cullen KE (2014) Loss of alpha-calcitonin gene-related peptide (alphaCGRP) reduces the efficacy of the Vestibulo-ocular Reflex (VOR). J Neurosci 34:10453–10458. https://doi.org/10.1523/JNEUROSCI.3336-13.2014

Maison SF, Emeson RB, Adams JC, Luebke AE, Liberman MC (2003) Loss of alpha CGRP reduces sound-evoked activity in the cochlear nerve. J Neurophysiol 90:2941–2949. https://doi.org/10.1152/jn.00596.2003

Marquez de Prado B, Hammond DL, Russo AF (2009) Genetic enhancement of calcitonin gene-related Peptide-induced central sensitization to mechanical stimuli in mice. J Pain 10:992–1000. https://doi.org/10.1016/j.jpain.2009.03.018

Marti E et al (1987) Ontogeny of peptide- and amine-containing neurones in motor, sensory, and autonomic regions of rat and human spinal cord, dorsal root ganglia, and rat skin. J Comp Neurol 266:332–359. https://doi.org/10.1002/cne.902660304

Martin VT, Behbehani MM (2001) Toward a rational understanding of migraine trigger factors. Med Clin North Am 85:911–941

Mason BN, Kaiser EA, Kuburas A, Loomis MM, Latham JA, Garcia-Martinez LF, Russo AF (2017) Induction of migraine-like photophobic behavior in mice by both peripheral and central cgrp mechanisms. J Neurosci 37:204–216. https://doi.org/10.1523/JNEUROSCI.2967-16.2016

Mathew NT, Kailasam J, Seifert T (2004) Clinical recognition of allodynia in migraine. Neurology 63:848–852

Melo-Carrillo A, Noseda R, Nir RR, Schain AJ, Stratton J, Strassman AM, Burstein R (2017a) Selective inhibition of trigeminovascular neurons by fremanezumab: a humanized monoclonal anti-CGRP. Antibody J Neurosci 37:7149–7163. https://doi.org/10.1523/JNEUROSCI.0576-17.2017

Melo-Carrillo A, Strassman AM, Nir RR, Schain AJ, Noseda R, Stratton J, Burstein R (2017b) Fremanezumab-A humanized monoclonal anti-CGRP antibody-inhibits thinly myelinated (adelta) but not unmyelinated (C) meningeal nociceptors. J Neurosci 37:10587–10596. https://doi.org/10.1523/JNEUROSCI.2211-17.2017

Messlinger K, Hanesch U, Kurosawa M, Pawlak M, Schmidt RF (1995) Calcitonin gene related peptide released from dural nerve fibers mediates increase of meningeal blood flow in the rat. Can J Physiol Pharmacol 73:1020–1024

Mitsikostas DD, Sanchez del Rio M, Moskowitz MA, Waeber C (1999) Both 5-HT1B and 5-HT1F receptors modulate c-fos expression within rat trigeminal nucleus caudalis. Eur J Pharmacol 369:271–277

Mogil JS et al (2005) Variable sensitivity to noxious heat is mediated by differential expression of the CGRP gene. Proc Natl Acad Sci U S A 102:12938–12943. https://doi.org/10.1073/pnas.0503264102

Mulleners WM, Aurora SK, Chronicle EP, Stewart R, Gopal S, Koehler PJ (2001) Self-reported photophobic symptoms in migraineurs and controls are reliable and predict diagnostic category accurately. Headache 41:31–39

Munro G, Petersen S, Jansen-Olesen I, Olesen J (2018) A unique inbred rat strain with sustained cephalic hypersensitivity as a model of chronic migraine-like pain. Sci Rep 8:1836. https://doi. org/10.1038/s41598-018-19901-1

Nakamura-Craig M, Gill BK (1991) Effect of neurokinin A, substance P and calcitonin gene related peptide in peripheral hyperalgesia in the rat paw. Neurosci Lett 124:49–51

Noseda R, Monconduit L, Constandil L, Chalus M, Villanueva L (2008) Central nervous system networks involved in the processing of meningeal and cutaneous inputs from the ophthalmic branch of the trigeminal nerve in the rat. Cephalalgia 28:813–824. https://doi.org/10.1111/j. 1468-2982.2008.01588.x

Oh-hashi Y et al (2001) Elevated sympathetic nervous activity in mice deficient in alphaCGRP. Circ Res 89:983–990

Oku R, Satoh M, Fujii N, Otaka A, Yajima H, Takagi H (1987) Calcitonin gene-related peptide promotes mechanical nociception by potentiating release of substance P from the spinal dorsal horn in rats. Brain Res 403:350–354

Ong JJY, Wei DY, Goadsby PJ (2018) Recent advances in pharmacotherapy for migraine prevention: from pathophysiology to new drugs. Drugs 78:411–437. https://doi.org/10.1007/s40265-018-0865-y

Oshinsky ML, Sanghvi MM, Maxwell CR, Gonzalez D, Spangenberg RJ, Cooper M, Silberstein SD (2012) Spontaneous trigeminal allodynia in rats: a model of primary headache. Headache 52:1336–1349. https://doi.org/10.1111/j.1526-4610.2012.02247.x

Pecile A, Guidobono F, Netti C, Sibilia V, Biella G, Braga PC (1987) Calcitonin gene-related peptide: antinociceptive activity in rats, comparison with calcitonin. Regul Pept 18:189–199

Petersen KA, Birk S, Doods H, Edvinsson L, Olesen J (2004) Inhibitory effect of BIBN4096BS on cephalic vasodilatation induced by CGRP or transcranial electrical stimulation in the rat. Br J Pharmacol 143:697–704. https://doi.org/10.1038/sj.bjp.0705966

Petersen KA, Lassen LH, Birk S, Lesko L, Olesen J (2005a) BIBN4096BS antagonizes human alpha-calcitonin gene related peptide-induced headache and extracerebral artery dilatation. Clin Pharmacol Ther 77:202–213

Petersen KA, Nilsson E, Olesen J, Edvinsson L (2005b) Presence and function of the calcitonin gene-related peptide receptor on rat pial arteries investigated in vitro and in vivo. Cephalalgia 25:424–432. https://doi.org/10.1111/j.1468-2982.2005.00869.x

Pradhan AA, Smith ML, McGuire B, Tarash I, Evans CJ, Charles A (2014) Characterization of a novel model of chronic migraine. Pain 155:269–274. https://doi.org/10.1016/j.pain.2013. 10.004

Rasmussen BK (1993) Migraine and tension-type headache in a general population: precipitating factors, female hormones, sleep pattern and relation to lifestyle. Pain 53:65–72

Rea BJ et al (2018) Peripherally administered calcitonin gene-related peptide induces spontaneous pain in mice: implications for migraine. Pain. https://doi.org/10.1097/j.pain.0000000000001337

Recober A, Kuburas A, Zhang Z, Wemmie JA, Anderson MG, Russo AF (2009) Role of calcitonin gene-related peptide in light-aversive behavior: implications for migraine. J Neurosci 29:8798–8804. https://doi.org/10.1523/JNEUROSCI.1727-09.2009

Recober A, Kaiser EA, Kuburas A, Russo AF (2010) Induction of multiple photophobic behaviors in a transgenic mouse sensitized to CGRP. Neuropharmacology 58:156–165. https://doi.org/10. 1016/j.neuropharm.2009.07.009

Reuter U et al (1998) Perivascular nerves contribute to cortical spreading depression-associated hyperemia in rats. Am J Physiol 274:H1979–H1987

Rosenfeld MG et al (1983) Production of a novel neuropeptide encoded by the calcitonin gene via tissue-specific RNA processing. Nature 304:129–135

Russell FA, King R, Smillie SJ, Kodji X, Brain SD (2014) Calcitonin gene-related peptide: physiology and pathophysiology. Physiol Rev 94:1099–1142. https://doi.org/10.1152/physrev.00034.2013

Russo AF (2015a) Calcitonin gene-related peptide (CGRP): a new target for migraine. Annu Rev Pharmacol Toxicol 55:533–552. https://doi.org/10.1146/annurev-pharmtox-010814-124701

Russo AF (2015b) CGRP as a neuropeptide in migraine: lessons from mice. Br J Clin Pharmacol 80:403–414. https://doi.org/10.1111/bcp.12686

Russo AF, Kuburas A, Kaiser EA, Raddant AC, Recober A (2009) A potential preclinical migraine model: CGRP-sensitized mice. Mol Cell Pharmacol 1:264–270

Sabharwal R, Zhang Z, Lu Y, Abboud FM, Russo AF, Chapleau MW (2010) Receptor activity-modifying protein 1 increases baroreflex sensitivity and attenuates angiotensin-induced hypertension. Hypertension 55:627–635. https://doi.org/10.1161/HYPERTENSIONAHA.109.148171

Salmon AM, Damaj I, Sekine S, Picciotto MR, Marubio L, Changeux JP (1999) Modulation of morphine analgesia in alphaCGRP mutant mice. Neuroreport 10:849–854

Salmon AM, Damaj MI, Marubio LM, Epping-Jordan MP, Merlo-Pich E, Changeux JP (2001) Altered neuroadaptation in opiate dependence and neurogenic inflammatory nociception in alpha CGRP-deficient mice. Nat Neurosci 4:357–358. https://doi.org/10.1038/86001

Satoh M, Kuraishi Y, Kawamura M (1992) Effects of intrathecal antibodies to substance P, calcitonin gene-related peptide and galanin on repeated cold stress-induced hyperalgesia: comparison with carrageenan-induced hyperalgesia. Pain 49:273–278

Schorscher-Petcu A, Austin JS, Mogil JS, Quirion R (2009) Role of central calcitonin gene-related peptide (CGRP) in locomotor and anxiety- and depression-like behaviors in two mouse strains exhibiting a CGRP-dependent difference in thermal pain sensitivity. J Mol Neurosci 39:125–136. https://doi.org/10.1007/s12031-009-9201-z

Senba E, Tohyama M (1988) Calcitonin gene-related peptide containing autonomic efferent pathways to the pelvic ganglia of the rat. Brain Res 449:386–390

Siren AL, Feuerstein G (1988) Cardiovascular effects of rat calcitonin gene-related peptide in the conscious rat. J Pharmacol Exp Ther 247:69–78

Skofitsch G, Jacobowitz DM (1985) Calcitonin gene-related peptide: detailed immunohistochemical distribution in the central nervous system. Peptides 6:721–745

Sowers LP, Tye AE, Russo AF (2017) Lessons learned from CGRP mutant mice. In: Dalkara T, Moskowitz MA (eds) Neurobiological basis of migraine, 1st edn. Wiley, Hoboken

Spierings EL, Ranke AH, Honkoop PC (2001) Precipitating and aggravating factors of migraine versus tension-type headache. Headache 41:554–558

Storer RJ, Akerman S, Goadsby PJ (2004) Calcitonin gene-related peptide (CGRP) modulates nociceptive trigeminovascular transmission in the cat. Br J Pharmacol 142:1171–1181. https://doi.org/10.1038/sj.bjp.0705807

Stucky NL, Gregory E, Winter MK, He YY, Hamilton ES, McCarson KE, Berman NE (2011) Sex differences in behavior and expression of CGRP-related genes in a rodent model of chronic migraine. Headache 51:674–692. https://doi.org/10.1111/j.1526-4610.2011.01882.x

Sun RQ, Lawand NB, Lin Q, Willis WD (2004) Role of calcitonin gene-related peptide in the sensitization of dorsal horn neurons to mechanical stimulation after intradermal injection of capsaicin. J Neurophysiol 92:320–326. https://doi.org/10.1152/jn.00086.2004

Tassorelli C, Greco R, Wang D, Sandrini M, Sandrini G, Nappi G (2003) Nitroglycerin induces hyperalgesia in rats – a time-course study. Eur J Pharmacol 464:159–162

Thomsen LL, Kruuse C, Iversen HK, Olesen J (1994) A nitric oxide donor (nitroglycerin) triggers genuine migraine attacks. Eur J Neurol 1:73–80. https://doi.org/10.1111/j.1468-1331.1994.tb00053.x

Troltzsch M, Denekas T, Messlinger K (2007) The calcitonin gene-related peptide (CGRP) receptor antagonist BIBN4096BS reduces neurogenic increases in dural blood flow. Eur J Pharmacol 562:103–110. https://doi.org/10.1016/j.ejphar.2007.01.058

Tsujikawa K et al (2007) Hypertension and dysregulated proinflammatory cytokine production in receptor activity-modifying protein 1-deficient mice. Proc Natl Acad Sci U S A 104:16702–16707. https://doi.org/10.1073/pnas.0705974104

Uddman R, Edvinsson L, Ekman R, Kingman T, McCulloch J (1985) Innervation of the feline cerebral vasculature by nerve fibers containing calcitonin gene-related peptide: trigeminal origin and co-existence with substance P. Neurosci Lett 62:131–136 van Dongen RM et al (2017) Migraine biomarkers in cerebrospinal fluid: a systematic review and meta-analysis. Cephalalgia 37:49–63. https://doi.org/10.1177/0333102415625614

Vos T et al (2012) Years lived with disability (YLDs) for 1160 sequelae of 289 diseases and injuries 1990-2010: a systematic analysis for the global burden of disease study 2010. Lancet 380:2163–2196. https://doi.org/10.1016/S0140-6736(12)61729-2

Wahl M, Schilling L, Parsons AA, Kaumann A (1994) Involvement of calcitonin gene-related peptide (CGRP) and nitric oxide (NO) in the pial artery dilatation elicited by cortical spreading depression. Brain Res 637:204–210

Wang Y, Li Y, Wang M (2016) Involvement of CGRP receptors in retinal spreading depression. Pharmacol Rep 68:935–938. https://doi.org/10.1016/j.pharep.2016.05.001

Weidner C et al (2000) Acute effects of substance P and calcitonin gene-related peptide in human skin – a microdialysis study. J Invest Dermatol 115:1015–1020. https://doi.org/10.1046/j.1523-1747.2000.00142.x

Wiesenfeld-Hallin Z, Hokfelt T, Lundberg JM, Forssmann WG, Reinecke M, Tschopp FA, Fischer JA (1984) Immunoreactive calcitonin gene-related peptide and substance P coexist in sensory neurons to the spinal cord and interact in spinal behavioral responses of the rat. Neurosci Lett 52:199–204

Williamson DJ, Hargreaves RJ, Hill RG, Shepheard SL (1997) Sumatriptan inhibits neurogenic vasodilation of dural blood vessels in the anaesthetized rat – intravital microscope studies. Cephalalgia 17:525–531. https://doi.org/10.1046/j.1468-2982.1997.1704525.x

Williamson DJ, Hill RG, Shepheard SL, Hargreaves RJ (2001) The anti-migraine 5-HT(1B/1D) agonist rizatriptan inhibits neurogenic dural vasodilation in anaesthetized guinea-pigs. Br J Pharmacol 133:1029–1034. https://doi.org/10.1038/sj.bjp.0704162

Xu W, Lundeberg T, Wang YT, Li Y, Yu LC (2003) Antinociceptive effect of calcitonin gene-related peptide in the central nucleus of amygdala: activating opioid receptors through amygdala-periaqueductal gray pathway. Neuroscience 118:1015–1022

Yao G, Huang Q, Wang M, Yang CL, Liu CF, Yu TM (2017) Behavioral study of a rat model of migraine induced by CGRP. Neurosci Lett 651:134–139. https://doi.org/10.1016/j.neulet.2017.04.059

Yisarakun W, Chantong C, Supornsilpchai W, Thongtan T, Srikiatkhachorn A, Reuangwechvorachai P, Maneesri-le Grand S (2015) Up-regulation of calcitonin gene-related peptide in trigeminal ganglion following chronic exposure to paracetamol in a CSD migraine animal model. Neuropeptides 51:9–16. https://doi.org/10.1016/j.npep.2015.03.008

Yu LC, Weng XH, Wang JW, Lundeberg T (2003) Involvement of calcitonin gene-related peptide and its receptor in anti-nociception in the periaqueductal grey of rats. Neurosci Lett 349:1–4

Zagami AS, Goadsby PJ, Edvinsson L (1990) Stimulation of the superior sagittal sinus in the cat causes release of vasoactive peptides. Neuropeptides 16:69–75

Zhang L, Hoff AO, Wimalawansa SJ, Cote GJ, Gagel RF, Westlund KN (2001) Arthritic calcitonin/alpha calcitonin gene-related peptide knockout mice have reduced nociceptive hypersensitivity. Pain 89:265–273

Zhang Z, Dickerson IM, Russo AF (2006) Calcitonin gene-related peptide receptor activation by receptor activity-modifying protein-1 gene transfer to vascular smooth muscle cells. Endocrinology 147:1932–1940. https://doi.org/10.1210/en.2005-0918

Zhang Z, Winborn CS, Marquez de Prado B, Russo AF (2007) Sensitization of calcitonin gene-related peptide receptors by receptor activity-modifying protein-1 in the trigeminal ganglion. J Neurosci 27:2693–2703. https://doi.org/10.1523/JNEUROSCI.4542-06.2007

Zhou X, Li JJ, Yu LC (2003) Plastic changes of calcitonin gene-related peptide in morphine tolerance: behavioral and immunohistochemical study in rats. J Neurosci Res 74:622–629. https://doi.org/10.1002/jnr.10770

CGRP in Human Models of Migraine

Håkan Ashina, Henrik Winther Schytz, and Messoud Ashina

Contents

Abstract

Over the past three decades, calcitonin gene-related peptide (CGRP) has emerged as a key molecule. Provocation experiments have demonstrated that intravenous CGRP infusion induces migraine-like attacks in migraine with and without aura patients. In addition, these studies have revealed a heterogeneous CGRP response, i.e., some migraine patients develop migraine-like attacks after CGRP infusion, while others do not. The role of CGRP in human migraine models has pointed to three potential sites of CGRP-induced migraine: (1) *vasodilation via cyclic adenosine monophosphate (cAMP) and possibly cyclic guanosine monophosphate (cGMP)*; (2) *activation of trigeminal sensory afferents,* and (3) *modulation of deep brain structures.* In the future, refined human experimental studies will continue to unveil the role of CGRP in migraine pathogenesis.

Keywords

CGRP · Human provocation models · Migraine

H. Ashina (✉) · H. W. Schytz · M. Ashina
Danish Headache Center, Department of Neurology, Rigshospitalet Glostrup, Faculty of Health and Medical Sciences, University of Copenhagen, Copenhagen, Denmark
e-mail: ashina@dadlnet.dk

S. D. Brain, P. Geppetti (eds.), *Calcitonin Gene-Related Peptide (CGRP) Mechanisms,*
Handbook of Experimental Pharmacology 255, https://doi.org/10.1007/164_2018_128

1   Introduction

CGRP is a key signaling molecule in migraine pathophysiology and anti-CGRP drugs constitute a new promising target in migraine treatment (Khan et al. 2017a). CGRP is widely distributed in the central and peripheral nervous system (Amara et al. 1982; Rosenfeld et al. 1983; van Rossum et al. 1997; Hostetler et al. 2013), including in the perivascular trigeminal sensory afferents (Uddman et al. 1985), trigeminal ganglion (Uddman et al. 1985), and trigeminal nucleus caudalis (TNC) (Uddman et al. 2002). Aside from its function as a potent dilator of human cerebral arteries (McCulloch et al. 1986; Edvinsson et al. 1987), CGRP is released upon activation of the trigeminal ganglion (Goadsby et al. 1988) and induces migraine-like attack in migraine patients (Lassen et al. 2002). Thus, it seems that CGRP plays an important role in the development of migraine attacks. This chapter reviews studies on CGRP-induced migraine and discusses possible mechanisms underlying migraine induction following CGRP infusion.

2   Methodology in Human Migraine Models

Human headache models have greatly improved our understanding of migraine pathophysiology (Ashina et al. 2013; Schytz et al. 2017). The first human headache model was developed through systematic research using intravenous infusion of the nitric oxide donor glyceryl trinitrate (GTN) (Iversen et al. 1989). It has been demonstrated that GTN provokes migraine attacks without aura in migraine patients with and without aura (Thomsen et al. 1994; Christiansen et al. 1999). To date, other human models have been developed including migraine provocation models using intravenous infusion of CGRP (Lassen et al. 2002) and PACAP38 (Schytz et al. 2009).

We will briefly discuss the main aspects to consider when applying the human migraine model. For a more detailed description on this, readers are referred to recently published work (Fig. 1) (Ashina et al. 2017). In general, human migraine models use a double-blinded, crossover design (Olesen et al. 2009), where patients are randomly allocated to receive intravenous infusion of pharmacological migraine "triggers" or placebo (isotonic saline). An 11-point numeric scale (NRS 11) from 0 to 10 (0 no headache, 10 worst imaginable headache) is used to record headache characteristics up to 24 h after the start of infusion and to evaluate whether the patients develop headache with typical migraine features. It is important to notice that only pharmacological signaling molecules that induce sufficient headache in healthy volunteers should be tested on migraine sufferers. Furthermore, healthy volunteers are used to determine an optimal dose causing vascular responses (e.g., dilation of cranial arteries) and headache. Provocation experiments in migraine patients require that patients are headache-free for at least 5 days before the experimental days. Therefore, patients with a moderate frequency of migraine attacks are preferable for recruitment. In addition, patients should not take preventive medication because it might influence the outcomes. Healthy volunteers with susceptibility

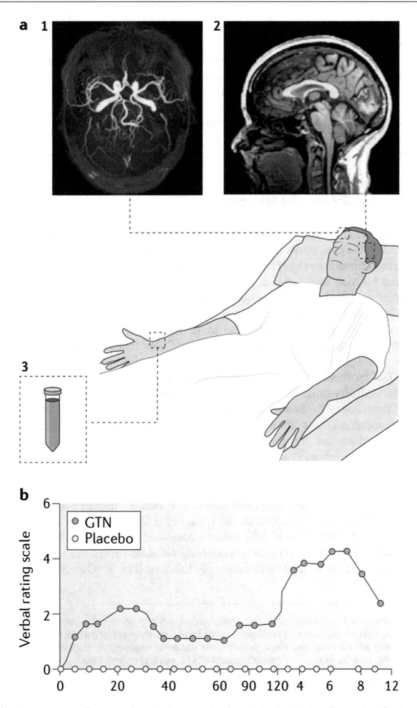

Fig. 1 The setup of an experimental human migraine study. (**a**) At baseline and at fixed and predefined intervals, the hemodynamic effects of the infusion are recorded, and might include

to migraine (first-degree relatives suffering from migraine) should be excluded to avoid triggering migraine.

Human provocation studies have also been combined with different recording techniques to investigate vascular biomarkers by transcranial Doppler ultrasonography (TCD) (Petersen et al. 2005; Lassen et al. 2008) and magnetic resonance imaging (MRI) angiography (Asghar et al. 2010, 2011) and CNS biomarkers (activation of deep brain structure) by functional MRI (fMRI) (Asghar et al. 2012, 2016) (Fig. 2).

3   CGRP-Induced Migraine

The first demonstration of CGRP-induced migraine-like attacks was documented in a double-blind crossover study (Lassen et al. 2002). Twelve patients suffering from migraine without aura (MO) were randomly allocated to receive CGRP (2 µg/min) or placebo infusion in the cubital vein over 20 min (Lassen et al. 2002). The final analyses excluded three of the patients, two of whom experienced hypotension associated with pallor and palpitations. The authors stated that the induced hypotension rendered the used CGRP dose to be the maximally tolerated dose. The remaining nine patients reported headache after CGRP infusion. All patients experienced flushing of the face, neck, and upper chest approximately 10 min after the start of CGRP infusion. Three patients reported migraine-like attacks according to the International Headache Society classification criteria for MO (Headache Classification Committee of the International Headache Society 1988). However, experimentally induced migraine attacks are not spontaneous and, therefore, we proposed different criteria for experimentally induced migraine (Hansen et al. 2010). By applying these criteria (Hansen et al. 2010) in the first CGRP provocation study (Lassen et al. 2002), we calculated that six out of nine patients (67%) experienced migraine-like attacks after CGRP infusion to only one after placebo. Later, two experimental studies also confirmed that CGRP induced migraine-like attacks in 67% of MO patients (Asghar et al. 2011; Guo et al. 2016). In addition, one of the studies reported that 40% of MO patients developed migraine-like attacks in the immediate phase (0–90 min post-infusion) (Guo et al. 2016) and the headache pain was predominantly frontally and temporally localized (Fig. 3) (Guo et al. 2016).

Fig. 1 (continued) recordings of the intracranial and extracranial arteries (1 – magnetic resonance angiography) or brain activity [2 – blood oxygen level-dependent functional MRI (fMRI)]. Vital signs, such as heart rate and blood pressure, are measured continuously throughout the study. Studies can be tailored to assess certain aspects – if the focus is to address imaging or plasma levels of a given substance, scans and blood sampling (3) are conducted at baseline, when effects are expected, and after treatment of the attack. **(b)** Headache intensity is recorded on a verbal rating scale from 0 to 10 (0, no headache; 5, moderate headache; 10, worst imaginable headache). Note the biphasic response, comprising an immediate headache followed hours later by a migraine-like headache (Ashina et al. 2017)

Fig. 2 Pie chart: percentage (numbers) of patients who developed delayed migraine-like attacks and patients who did not develop delayed migraine-like attacks after CGRP infusion (Lassen et al. 2002; Asghar et al. 2011; Guo et al. 2016; Hansen et al. 2010)

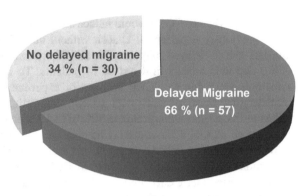

Fig. 3 Usual (●/red %) and CGRP-induced (●/blue %) localization of migraine attacks (Guo et al. 2016)

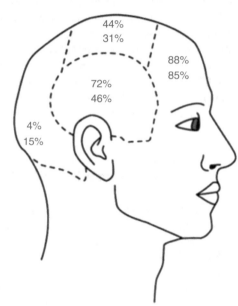

The role of CGRP in migraine with aura (MA) is not fully clarified (Hansen and Ashina 2014). Cortical spreading depression (CSD), a slowly propagated wave of depolarization followed by suppression of brain activity, is likely the underlying mechanism of the migraine aura (Charles and Baca 2013). In rats, endogenous CGRP was released from cortical slices and CGRP receptor antagonists had a dose-dependent inhibitory effect on CSD (Tozzi et al. 2012). In cats, CSD had no effect on the concentration of CGRP in the external jugular vein (Piper et al. 1993). One study investigated plasma CGRP collected from the carotid artery and the internal jugular vein in conjunction with cerebral angiography in MA patients (Friberg et al. 1994). Repeated measurements of plasma CGRP levels during MA attacks revealed no changes over time in arterial-venous plasma concentrations or in the release rates of CGRP (Friberg et al. 1994). Moreover, one provocation study investigated whether CGRP would induce aura attacks in patients with MA who had *never* previously experienced attacks without aura (Hansen et al. 2010). CGRP infusion caused delayed migraine-like attacks *without* aura in 8 out of 14 patients (57%) and aura attacks in 4 out of 14 patients. To investigate whether the aura symptoms were due to CGRP infusion or experimental stress, it would require rechallenging these patients with CGRP. Interestingly, GTN infusion induces migraine-like attacks *without* aura in 50% of MA patients who have never previously experienced attacks *without* aura (Christiansen et al. 1999). In relation to familial hemiplegic migraine (FHM), an autosomal dominant subtype of MA, two provocation studies investigated the response to CGRP infusion in FHM patients. Both studies reported that CGRP infusion did not induce migraine-like attacks in FHM patients with (Hansen et al. 2008) and without known mutations (Hansen et al. 2011). These findings suggested that neurobiological pathways responsible for migraine headache in MO and MA patients may be distinct from pathways responsible for migraine headache in FHM patients (Hansen et al. 2008).

Taken together, studies in patients with common types of migraine demonstrated that CGRP infusion caused delayed migraine-like attacks in 66% of migraine patients with and without aura (Fig. 2) (Lassen et al. 2002; Asghar et al. 2011; Guo et al. 2016; Hansen et al. 2010).

4   Possible Mechanisms of CGRP-Induced Migraine

Potential sites of CGRP-induced migraine include *vasodilation* via *cyclic adenosine monophosphate (cAMP) and possibly cyclic guanosine monophosphate (cGMP)*, *peripheral activation of trigeminal sensory afferents*, and *central neuromodulation of higher brain centers*. Of these possible mechanisms of action, the peripheral vascular effects of CGRP have been the most extensively studied aspect. CGRP binds to its receptor on smooth muscle cells and acts as a potent dilator of human cerebral arteries (Edvinsson et al. 1987). In both healthy volunteers (Asghar et al. 2010) and MO patients (Asghar et al. 2011), MRA studies have demonstrated CGRP-induced dilation in the extracranial part of the middle meningeal artery (MMA), which was reversed following sumatriptan administration. The question then arises as to what extent CGRP-induced arterial dilation contributes to provoked migraine attacks. Studies have hypothesized that cAMP and possibly cGMP might be upregulated intracellularly following CGRP's extracellular binding to its receptor on vascular smooth muscle cells (Edvinsson et al. 1985; Uddman et al. 1985). Cilostazol, which induces cAMP elevation in vascular smooth muscle cells via inhibition of phosphodiesterase 3 dependent degradation, provokes migraine attacks in 86% of patients (Guo et al. 2014; Khan et al. 2017b). Nitric oxide production regulates cGMP formation and the nitric oxide donor GTN and sildenafil (a highly selective inhibitor of phosphodiesterase 5 that breaks down cGMP, and its inhibition leads to accumulation of cGMP) are powerful migraine triggers (Tvedskov et al. 2010; Kruuse et al. 2003).

To explore the possible relationship between nitric oxide and CGRP signaling pathways, one study investigated the effect of a nitric oxide-synthase inhibitor N(G)-monomethyl-L-arginine (L-NMMA) on CGRP-induced vasodilation (de Hoon et al. 2003). In a forearm skin model, L-NMMA infusion reduced CGRP-induced vasodilation in 40 healthy volunteers. Interestingly, L-NMMA did not have an inhibitory effect on CGRP-induced vasodilation when the highest CGRP dose was used. One provocation study investigated the effect of the CGRP receptor antagonist olcegepant in prevention of GTN-induced migraine (Tvedskov et al. 2010). In crossover fashion, all participants were pre-treated with olcegepant or placebo followed by infusion of GTN. This study showed no effect of olcegepant in prevention of GTN-induced migraine (Tvedskov et al. 2010). Collectively, these data suggest that activation of CGRP signaling pathway through intracellular cGMP increase is unlikely to be involved in mechanisms underlying CGRP-induced migraine.

One of the important questions that should be addressed is if CGRP may directly act on the trigeminal sensory afferents. An in vitro study reported CGRP release from capsaicin sensitive nerve fibers and dilation of human cerebral arteries (Jansen-Olesen et al. 1996). In rats, CGRP-induced meningeal vasodilation did not activate or sensitize meningeal nociceptors (Levy et al. 2005). In humans, CGRP injection into the forearm skin did not elicit a pain sensation (Pedersen-Bjergaard et al. 1991). CGRP receptors are expressed in three levels of the trigeminovascular system (Lennerz et al. 2008): peripheral nerve fibers associated with the cranial dura mater, the trigeminal ganglion, and the spinal trigeminal nucleus. Theoretically, exogenously administered CGRP could act on these receptors and induce migraine. In support, CGRP activates transcription of pro-nociceptive receptors on cultured trigeminal neurons through protein kinase A-dependent mechanisms (Giniatullin et al. 2008). In humans, noxious heat stimuli applied to the V1 area of the trigeminal nerve after CGRP infusion caused blood oxygen level-dependent (BOLD), a surrogate marker of neuronal activity, changes in the insula, brain stem, caudate nuclei, thalamus, and cingulate cortex that were reversed by administration of sumatriptan (Asghar et al. 2016). Given that neither CGRP nor sumatriptan are likely to cross the blood–brain barrier (BBB), these data indicated that CGRP might modulate nociceptive trigeminal transmission without having a direct effect in the CNS (Asghar et al. 2016). Furthermore, an fMRI study showed that visual sensory input by checkerboard stimulation did not cause any BOLD signal changes in the visual cortex after CGRP infusion (Asghar et al. 2012). In addition, none of the CGRP provocation studies reported any CNS side effects after CGRP infusion (Lassen et al. 2002; Asghar et al. 2011; Guo et al. 2016; Hansen et al. 2010). Collectively, these data support the notion of a predominantly peripheral site of action of CGRP in migraine. A preclinical study on anesthetized rats provided further support for peripheral mechanism by demonstrating that a monoclonal anti-CGRP antibody, fremanezumab, inhibited naive high-threshold neurons, but not wide-dynamic range trigeminovascular neurons (Melo-Carrillo et al. 2017a). Moreover, the inhibitory effects on the neurons were limited to their activation from the intracranial dura but *not* facial skin or cornea (Melo-Carrillo et al. 2017a). Interestingly, fremanezumab inhibited the meningeal nociceptors of Aδ-fiber neurons but not C-fiber neurons (Melo-Carrillo et al. 2017b). The importance of these studies (Melo-Carrillo et al. 2017a, b) is twofold. First, fremanezumab selectively inhibited meningeal nociceptors of Aδ-fiber neurons peripherally and naive high-threshold trigeminovascular neurons centrally. Therefore, it has been suggested (Melo-Carrillo et al. 2017a, b) that the headache generating phase of migraine relies on activation of meningeal nociceptors because the site of action for monoclonal anti-CGPR antibodies is situated outside the BBB. Secondly, it has been proposed (Melo-Carrillo et al. 2017a, b) a possible explanation to why monoclonal anti-CGRP antibodies are not effective in all migraine patients in that fremanezumab only exhibited selective inhibition of meningeal nociceptors. Thus, it could be speculated that inhibition of meningeal nociceptors of Aδ-fiber neurons and naive high-threshold trigeminovascular neurons is not sufficient to block the generation of a migraine attack in all patients.

The question is whether CGRP could induce migraine via centrally mediated mechanisms? In this context, we should point to arguments favoring a central site of action for CGRP in migraine pathogenesis: (1) CGRP is widely distributed in central migraine-relevant structures such as the TNC and the spinal cord at the C1-level (Eftekhari and Edvinsson 2011); (2) Gene expression of the CGRP receptor has been documented within the spinal trigeminal nucleus, pons, and spinal cord (Eftekhari et al. 2016); (3) CGRP could also act on second-order neurons in the trigeminocervical complex (TCC) leading to nociceptive transmission to the thalamus and higher brain centers (Akerman et al. 2011). CGRP has also been implicated in modulation of central nociceptive transmission in the dorsal horn of the spinal cord (Yu et al. 2002). In addition, one study reported that olcegepant, CGRP receptor antagonist, inhibited trigeminovascular nociceptive transmission when injected into the periaqueductal gray of the midbrain (Pozo-Rosich et al. 2015).

To study the possible CNS-mediated migraine inducing effects, one study investigated premonitory symptoms as a surrogate for CNS involvement before onset of migraine headache (Guo et al. 2016). This study found that MO attacks were not associated with premonitory symptoms – suggesting a peripheral origin of CGRP-induced migraine. In support, one in vivo positron emission tomography (PET) study demonstrated that telcagepant, a CGRP receptor antagonist, at an efficacious dose only achieved low human CGRP receptor occupancy (Hostetler et al. 2013) – indicating that inhibition of central CGRP receptors is not needed for migraine amelioration. Furthermore, the PET tracer concentration was highest in the cerebellum in both anesthetized rhesus monkeys and healthy male volunteers – consistent with the known CGRP receptor distribution. Taken together, these data suggest a peripheral mechanism of action in CGRP-induced migraine.

5 Why Are Human Migraine Models Important?

Migraine is characterized by multiphasic events and its key feature is that it can be provoked by various triggers. Provocation models allow us to study migraine during different attack phases, i.e., from the beginning to the end. The clinical phenotype of pharmacologically induced migraine-like attacks mimics those of spontaneous migraine attacks. Thus, the unique value of human migraine models has been twofold.

First, it has led to the discovery of novel signaling pathways implicated in cascade of events that leads to a migraine attack. Studies have documented the importance of signaling peptides such as CGRP, GTN, and PACAP in inducing migraine-like attacks. In this context, the potential role of the cAMP and cGMP signaling is supported by provocation studies demonstrating that phosphodiesterase inhibitors induce migraine-like attacks with an aptitude superior to CGRP, GTN, and PACAP (Guo et al. 2014; Kruuse et al. 2003; Thomsen et al. 1994; Lassen et al. 2002; Schytz et al. 2009). The disparity suggests that phosphodiesterase inhibitors likely act downstream from the abovementioned signaling peptides in the migraine-generating cascade.

Secondly, studies on migraine provocation by CGRP provided strong support for development of anti-CGRP drugs for the treatment of migraine. Currently, several randomized clinical trials have demonstrated efficacy of anti-CGRP monoclonal antibodies in migraine prevention (Khan et al. 2017a). The CGRP provocation studies have also suggested the heterogenic CGRP response of migraine patients (Lassen et al. 2002; Asghar et al. 2011; Guo et al. 2016; Hansen et al. 2010), i.e., some patients developed attacks while others did not. Thus, the human migraine model could theoretically be used for stratifying migraine patients in CGRP responders and nonresponders in order to predict the response to anti-CGRP drugs. However, it should be noted that the lack of CGRP response has not been reproduced in the same group of nonresponders. Reproducibility is important to minimize the risk of fluctuating susceptibility to migraine, which could theoretically affect the CGRP response. Moreover, CGRP retest reproducibility would cement the value of CGRP as a migraine-specific biomarker.

6   Future Perspectives and Concluding Remarks

Future studies should seek to optimize study designs and explore unanswered questions concerning CGRP-induced migraine. One pivotal aspect would be to investigate possible innate fluctuations in migraine attack susceptibility. This could be addressed by examining the migraine provoking effect of CGRP in patients whose frequency of attacks has been documented for an extended period before and after infusion day. For this purpose, it would be interesting to investigate the CGRP response in migraine sufferers who only experience attacks very few times on a yearly basis. This would allow us to evaluate whether headache frequency impacts the likelihood of developing migraine-like attacks after CGRP infusion. Another approach would be to rechallenge a population of CGRP nonresponders and examine whether the lack of response is reproducible. If so, this would delineate CGRP as potent and reliable inducer of migraine attacks. In this context, future studies should also examine whether the efficacy of anti-CGRP antibody treatment can be predicted in migraine sufferers based on hypersensitivity to CGRP.

Human migraine models provide a unique opportunity to study migraine-specific mechanisms after CGRP infusion. The knowledge acquired from these studies has validated CGRP as potent inducer of migraine, thereby, paving the way for development of drugs targeting the CGRP molecule or its receptor. In the future, refined human experimental studies will continue to unveil the role of CGRP in migraine pathogenesis.

References

Akerman S, Holland PR, Goadsby PJ (2011) Diencephalic and brainstem mechanisms in migraine. Nat Rev Neurosci 12(10):570–584

Amara SG, Jonas V, Rosenfeld MG, Ong ES, Evans RM (1982) Alternative RNA processing in calcitonin gene expression generates mRNAs encoding different polypeptide products. Nature 298(5871):240–244

Asghar MS, Hansen AE, Kapijimpanga T, van der Geest RJ, van der Koning P, Larsson HB, Olesen J, Ashina M (2010) Dilation by CGRP of middle meningeal artery and reversal by sumatriptan in normal volunteers. Neurology 75(17):1520–1526

Asghar MS, Hansen AE, Amin FM, van der Geest RJ, Koning PV, Larsson HB, Olesen J, Ashina M (2011) Evidence for a vascular factor in migraine. Ann Neurol 69(4):635–645

Asghar MS, Hansen AE, Larsson HB, Olesen J, Ashina M (2012) Effect of CGRP and sumatriptan on the BOLD response in visual cortex. J Headache Pain 13(2):159–166

Asghar MS, Becerra L, Larsson HB, Borsook D, Ashina M (2016) Calcitonin gene-related peptide modulates heat nociception in the human brain – an fMRI study in healthy volunteers. PLoS One 11(3):e0150334

Ashina M, Hansen JM, Olesen J (2013) Pearls and pitfalls in human pharmacological models of migraine: 30 years' experience. Cephalalgia 33(8):540–553

Ashina M, Hansen JM, Á Dunga BO, Olesen J (2017) Human models of migraine – short-term pain for long-term gain. Nat Rev Neurol 13(12):713–724

Charles AC, Baca SM (2013) Cortical spreading depression and migraine. Nat Rev Neurol 9(11):637–644

Christiansen I, Thomsen LL, Daugaard D, Ulrich V, Olesen J (1999) Glyceryl trinitrate induces attacks of migraine without aura in sufferers of migraine with aura. Cephalalgia 19(7):660–667

de Hoon JN, Pickkers P, Smits P, Struijker-Boudier HA, Van Bortel LM (2003) Calcitonin gene-related peptide: exploring its vasodilating mechanism of action in humans. Clin Pharmacol Ther 73(4):312–321

Edvinsson L, Fredholm BB, Hamel E, Jansen I, Verrecchia C (1985) Perivascular peptides relax cerebral arteries concomitant with stimulation of cyclic adenosine monophosphate accumulation or release of an endothelium-derived relaxing factor in the cat. Neurosci Lett 58(2):213–217

Edvinsson L, Ekman R, Jansen I, McCulloch J, Uddman R (1987) Calcitonin gene-related peptide and cerebral blood vessels: distribution and vasomotor effects. J Cereb Blood Flow Metab 7(6):720–728

Eftekhari S, Edvinsson L (2011) Calcitonin gene-related peptide (CGRP) and its receptor components in human and rat spinal trigeminal nucleus and spinal cord at C1-level. BMC Neurosci 12:112

Eftekhari S, Gaspar RC, Roberts R, Chen TB, Zeng Z, Villarreal S, Edvinsson L, Salvatore CA (2016) Localization of CGRP receptor components and receptor binding sites in rhesus monkey brainstem: a detailed study using in situ hybridization, immunofluorescence, and autoradiography. J Comp Neurol 524(1):90–118

Friberg L, Olesen J, Olsen TS, Karle A, Ekman R, Fahrenkrug J (1994) Absence of vasoactive peptide release from brain to cerebral circulation during onset of migraine with aura. Cephalalgia 14(1):47–54

Giniatullin R, Nistri A, Fabbretti E (2008) Molecular mechanisms of sensitization of pain-transducing P2X3 receptors by the migraine mediators CGRP and NGF. Mol Neurobiol 37(1):83–90

Goadsby PJ, Edvinsson L, Ekman R (1988) Release of vasoactive peptides in the extracerebral circulation of humans and the cat during activation of the trigeminovascular system. Ann Neurol 23(2):193–196

Guo S, Olesen J, Ashina M (2014) Phosphodiesterase 3 inhibitor cilostazol induces migraine-like attacks via cyclic AMP increase. Brain 137(Pt 11):2951–2959

Guo S, Vollesen AL, Olesen J, Ashina M (2016) Premonitory and nonheadache symptoms induced by CGRP and PACAP38 in patients with migraine. Pain 157(12):2773–2781

Hansen JM, Ashina M (2014) Calcitonin gene-related peptide and migraine with aura: a systematic review. Cephalalgia 34(9):695–707

Hansen JM, Thomsen LL, Olesen J, Ashina M (2008) Calcitonin gene-related peptide does not cause the familial hemiplegic migraine phenotype. Neurology 71(11):841–847

Hansen JM, Hauge AW, Olesen J, Ashina M (2010) Calcitonin gene-related peptide triggers migraine-like attacks in patients with migraine with aura. Cephalalgia 30(10):1179–1186

Hansen JM, Thomsen LL, Olesen J, Ashina M (2011) Calcitonin gene-related peptide does not cause migraine attacks in patients with familial hemiplegic migraine. Headache 51(4):544–553

Headache Classification Committee of the International Headache Society (1988) Classification and diagnostic criteria for headache disorders, cranial neuralgias and facial pain. Cephalalgia 8(Suppl 7):1–96

Hostetler ED, Joshi AD, Sanabria-Bohórquez S, Fan H, Zeng Z, Purcell M, Gantert L, Riffel K, Williams M, O'Malley S, Miller P, Selnick HG, Gallicchio SN, Bell IM, Salvatore CA, Kane SA, Li CC, Hargreaves RJ, de Groot T, Bormans G, Van Hecken A, Derdelinckx I, de Hoon J, Reynders T, Declercq R, De Lepeleire I, Kennedy WP, Blanchard R, Marcantonio EE, Sur C, Cook JJ, Van Laere K, Evelhoch JL (2013) In vivo quantification of calcitonin gene-related peptide (CGRP) receptor occupancy by telcagepant in rhesus monkey and human brain using the positron emission tomography (PET) tracer [11C]MK-4232. J Pharmacol Exp Ther 347(2):478–486

Iversen HK, Olesen J, Tfelt-Hansen P (1989) Intravenous nitroglycerin as an experimental model of vascular headache. Basic characteristics. Pain 38(1):17–24

Jansen-Olesen I, Mortensen A, Edvinsson L (1996) Calcitonin gene-related peptide is released from capsaicin-sensitive nerve fibres and induces vasodilatation of human cerebral arteries concomitant with activation of adenylyl cyclase. Cephalalgia 16(5):310–316

Khan S, Deen M, Hougaard A, Amin FM, Ashina M (2017a) Reproducibility of migraine-like attacks induced by phosphodiesterase-3-inhibitor cilostazol. Cephalalgia. https://doi.org/10.1177/0333102417719753

Khan S, Olesen A, Ashina M (2017b) CGRP, a target for preventive therapy in migraine and cluster headache: systematic review of clinical data. Cephalalgia. https://doi.org/10.1177/0333102417741297

Kruuse C, Thomsen LL, Birk S, Olesen J (2003) Migraine can be induced by sildenafil without changes in middle cerebral artery diameter. Brain 126(Pt 1):241–247

Lassen LH, Haderslev PA, Jacobsen VB, Iversen HK, Sperling B, Olesen J (2002) CGRP may play a causative role in migraine. Cephalalgia 22(1):54–61

Lassen LH, Jacobsen VB, Haderslev PA, Sperling B, Iversen HK, Olesen J, Tfelt-Hansen P (2008) Involvement of calcitonin gene-related peptide in migraine: regional cerebral blood flow and blood flow velocity in migraine patients. J Headache Pain 9(3):151–157

Lennerz JK, Rühle V, Ceppa EP, Neuhuber WL, Bunnett NW, Grady EF, Messlinger K (2008) Calcitonin receptor-like receptor (CLR), receptor activity-modifying protein 1 (RAMP1), and calcitonin gene-related peptide (CGRP) immunoreactivity in the rat trigeminovascular system: differences between peripheral and central CGRP receptor distribution. J Comp Neurol 507(3):1277–1299

Levy D, Burstein R, Strassman AM (2005) Calcitonin gene-related peptide does not excite or sensitize meningeal nociceptors: implications for the pathophysiology of migraine. Ann Neurol 58(5):698–705

McCulloch J, Uddman R, Kingman TA, Edvinsson L (1986) Calcitonin gene-related peptide: functional role in cerebrovascular regulation. Proc Natl Acad Sci U S A 83(15):5731–5735

Melo-Carrillo A, Noseda R, Nir RR, Schain AJ, Stratton J, Strassman AM, Burstein R (2017a) Selective inhibition of trigeminovascular neurons by fremanezumab: a humanized monoclonal anti-CGRP antibody. J Neurosci 37(30):7149–7163

Melo-Carrillo A, Strassman AM, Nir RR, Schain AJ, Noseda R, Stratton J, Burstein R (2017b) Fremanezumab – a humanized monoclonal anti-CGRP antibody – inhibits thinly myelinated (Aδ) but not unmyelinated (C) meningeal nociceptors. J Neurosci 37(44):10587–10596

Olesen J, Tfelt-Hansen P, Ashina M (2009) Finding new drug targets for the treatment of migraine attacks. Cephalalgia 29(9):909–920

Pedersen-Bjergaard U, Nielsen LB, Jensen K, Edvinsson L, Jansen I, Olesen J (1991) Calcitonin gene-related peptide, neurokinin A and substance P: effects on nociception and neurogenic inflammation in human skin and temporal muscle. Peptides 12(2):333–337

Petersen KA, Lassen LH, Birk S, Lesko L, Olesen J (2005) BIBN4096BS antagonizes human alpha-calcitonin gene related peptide-induced headache and extracerebral artery. Clin Pharmacol Ther 77(3):202–213

Piper RD, Edvinsson L, Ekman R, Lambert GA (1993) Cortical spreading depression does not result in the release of calcitonin gene-related peptide into the external jugular vein of the cat: relevance to human migraine. Cephalalgia 13(3):180–183

Pozo-Rosich P, Storer RJ, Charbit AR, Goadsby PJ (2015) Periaqueductal gray calcitonin gene-related peptide modulates trigeminovascular neurons. Cephalalgia 35(14):1298–1307

Rosenfeld MG, Mermod JJ, Amara SG, Swanson LW, Sawchenko PE, Rivier J, Vale WW, Evans RM (1983) Production of a novel neuropeptide encoded by the calcitonin gene via tissue-specific RNA processing. Nature 304(5922):129–135

Schytz HW, Birk S, Wienecke T, Kruuse C, Olesen J, Ashina M (2009) PACAP38 induces migraine-like attacks in patients with migraine without aura. Brain 132(Pt 1):16–25

Schytz HW, Hargreaves R, Ashina M (2017) Challenges in developing drugs for primary headaches. Prog Neurobiol 152:70–88

Thomsen LL, Kruuse C, Iversen HK, Olesen J (1994) A nitric oxide donor (nitroglycerin) triggers genuine migraine attacks. Eur J Neurol 1(1):73–80

Tozzi A, de Iure A, Di Filippo M, Costa C, Caproni S, Pisani A, Bonsi P, Picconi B, Cupini LM, Materazzi S, Geppetti P, Sarchielli P, Calabresi P (2012) Critical role of calcitonin gene-related peptide receptors in cortical spreading depression. Proc Natl Acad Sci U S A 109 (46):18985–18990

Tvedskov JF, Tfelt-Hansen P, Petersen KA, Jensen LT, Olesen J (2010) CGRP receptor antagonist olcegepant (BIBN4096BS) does not prevent glyceryl trinitrate-induced migraine. Cephalalgia 30(11):1346–1353

Uddman R, Edvinsson L, Ekman R, Kingman T, McCulloch J (1985) Innervation of the feline cerebral vasculature by nerve fibers containing calcitonin gene-related peptide: trigeminal origin and co-existence with substance P. Neurosci Lett 62(1):131–136

Uddman R, Tajti J, Hou M, Sundler F, Edvinsson L (2002) Neuropeptide expression in the human trigeminal nucleus caudalis and in the cervical spinal cord C1 and C2. Cephalalgia 22 (2):112–116

van Rossum D, Hanisch UK, Quirion R (1997) Neuroanatomical localization, pharmacological characterization and functions of CGRP, related peptides and their receptors. Neurosci Biobehav Rev 21(5):649–678

Yu Y, Lundeberg T, Yu LC (2002) Role of calcitonin gene-related peptide and its antagonist on the evoked discharge frequency of wide dynamic range neurons in the dorsal horn of the spinal cord in rats. Regul Pept 103(1):23–27

Role of CGRP in Migraine

Lars Edvinsson

Contents

Abstract

Migraine is a common neurological disorder that afflicts up to 15% of the adult population in most countries, with predominance in females. It is characterized by episodic, often disabling headache, photophobia and phonophobia, autonomic symptoms (nausea and vomiting), and in a subgroup an aura in the beginning of the attack. Although still debated, many researchers consider migraine to be a disorder in which CNS dysfunction plays a pivotal role while various parts of the trigeminal system are necessary for the expression of associated symptoms.

Treatment of migraine has in recent years seen the development of drugs that target the trigeminal sensory neuropeptide calcitonin gene-related peptide (CGRP) or its receptor. Several of these drugs are now approved for use in frequent episodic and in chronic migraine. CGRP-related therapies offer considerable improvements over existing drugs, as they are the first to be designed

L. Edvinsson (✉)
Division of Experimental Vascular Research, Department of Clinical Sciences, Lund University, Lund, Sweden

Department of Clinical Experimental Research, Glostrup Research Institute, Glostrup Hospital, Glostrup, Denmark
e-mail: lars.edvinsson@med.lu.se

© Springer Nature Switzerland AG 2019   121
S. D. Brain, P. Geppetti (eds.), *Calcitonin Gene-Related Peptide (CGRP) Mechanisms*,
Handbook of Experimental Pharmacology 255, https://doi.org/10.1007/164_2018_201

specifically to act on the trigeminal pain system: they are more specific and have little or no adverse effects. Small molecule CGRP receptor antagonists, gepants, are effective for acute relief of migraine headache, whereas monoclonal antibodies against CGRP (Eptinezumab, Fremanezumab, and Galcanezumab) or the CGRP receptor (Erenumab) effectively prevent migraine attacks. The neurobiology of CGRP signaling is briefly summarized together with key clinical evidence for the role of CGRP in migraine headache, including the efficacy of CGRP-targeted treatments.

Keywords

CGRP · CGRP receptor · Gepants · Migraine · Monoclonal antibodies

1   Background

Migraine is a neurological disorder that afflicts up to 15% of the adult population, with predominance in females. It is a chronic, complex neurological disorder that manifests as recurrent attacks of moderate to severe headache pain lasting 4–72 h. It is characterized by episodic headache, photophobia and phonophobia, autonomic symptoms (nausea and vomiting), and in a subgroup an aura in the beginning of the attack (IHS 2018 Classification). Although still debated, nowadays migraine is considered a neurological disorder in which CNS dysfunction plays a pivotal role while various parts of the trigeminal system are necessary for the expression of peripheral associated symptoms.

In an early study, Weiller et al. (1995) suggested a "migraine generator region" in the brainstem. This has recently been revisited and a series of imaging studies have suggested that already on the day before the start of a migraine attack signs of activation in the hypothalamus, possibly involving thalamus and the limbic system, are present (May and Schulte 2016). Further connectivity imaging studies suggested that once in the attack brainstem regions and the trigeminovascular pathway are recruited eliciting many of the components classically linked to the symptomatology of the migraine attack (May 2017). The progression of the disease into different phases has nicely been elucidated recently (Charles 2018; Dodick 2018).

The trigeminal system is by and large involved in the pain part of the attack exemplified by the release of calcitonin gene-related peptide (CGRP) in the headache phase of the migraine attack and its termination by triptan administration (Goadsby and Edvinsson 1993; Goadsby et al. 1990). Interestingly, about half of all neurons in the trigeminal ganglion express CGRP, as visualized using immunohistochemical staining with CGRP antibodies (Edvinsson et al. 2018). Neurobiology studies have so far only shown CGRP to be directly linked to the attack, but reasonably other neuronal messengers could, in addition, be involved. Recent findings provide new insight into the intricate role of CNS and the trigeminal system in the pathophysiology of migraine. Current specific migraine therapies have been found to interact with the trigeminal system to limit symptoms of the migraine attack (Edvinsson et al. 2018).

2 CGRP in the Cranial System

The expression of mRNA from the calcitonin gene is tissue specific in which CGRP mRNA is predominantly expressed in nerves and calcitonin mRNA in the thyroid (Russell et al. 2014). The 37 amino acid peptide CGRP and calcitonin belong to a family of related molecules that include the peptides adrenomedullin, primarily produced by non-neuronal tissues, especially vascular tissues, and amylin that is mainly produced in the pancreas. They share some structural homology (approximately 25–40%) and there are also some similarities in biological activities (Hay et al. 2018). CGRP is abundant in the body and has a wide distribution throughout the central and peripheral nervous systems (Goadsby et al. 2017). CGRP is an extremely potent and long-lasting vasodilator that is active at all levels of the cardiovascular system (Uddman et al. 1985, 1986a). It is now realized that CGRP is a widely expressed neuropeptide that has a major role in sensory neurotransmission.

Soon after CGRP was discovered in 1982 (Amara et al. 1982), it was linked to the trigeminovascular system with implications for migraine. Our interest in this peptide began due to a general interest in neuronal messengers in autonomic and sensory innervation of the cranial circulation. We had already made progress upon challenging the classical dogma that "one nerve only has one neuronal messenger" (law of Canon) (Edvinsson and Uddman 2005) and could verify that neurons in the trigeminal ganglion co-localized CGRP, substance P, neurokinin A, neuronal nitric oxide synthase, pituitary adenylate cyclase activating peptide (PACAP), inter alia (Edvinsson and Uddman 2005). However, their individual and collective functional roles were still unknown.

The first data on CGRP in the trigeminovascular system were presented in 1984 at meeting on Regulatory Peptides (Edvinsson 1985) and later in a series of publications. In the animal and human trigeminal ganglion, over half of the neurons contain CGRP (Uddman et al. 1986b; Tajti et al. 1999). This was confirmed in subsequent work where two main populations of neurons in the trigeminal ganglion were quantified in several species including man (Eftekhari et al. 2010): half of them contain CGRP (C-fibers) and half the CGRP receptor elements (Aδ-fibers). In addition, >90% of the cell count consisted of small satellite glial cells organized around the neurons. The origin of the CGRP containing nerve fibers was demonstrated by immunohistochemistry and quantitation by radioimmunoassay, following trigeminal nerve denervation (Uddman et al. 1985). Subsequent tracing studies from intra- and extracranial arteries revealed that the sensory CGRP-positive fibers originate in the trigeminal neurons and co-localize with substance P (Edvinsson et al. 1989; Hara et al. 1989; Uddman and Edvinsson 1989).

There exist two forms of CGRP in the body; the α-CGRP is encoded by the CT/CGRP gene that is relevant for the cerebral circulation and migraine. The β-CGRP (with a structural homology of 90%) is primarily found in the gut and formed by a different gene. The α-CGRP is synthesized in neurons by tissue-specific splicing of mRNA transcribed from the calcitonin/CGRP gene located on chromosome 11. CGRP is generated by cleavage of a pro-peptide precursor, and then processed through the Golgi apparatus, packaged into dense core vesicles for transport to axon terminals for storage and release.

Functional studies showed that CGRP is a very potent vasodilator of cerebral arteries and arterioles, activating adenylyl cyclase in the smooth muscle cells and is unrelated to functional endothelium (Edvinsson 1985). In vivo, CGRP potently relaxed cortical arterioles but not venules (McCulloch et al. 1986). Lesioning of the perivascular sensory CGRP nerves, however, did not modify resting cerebral blood flow, flow-metabolism coupling, chemical regulation, or autoregulation of brain circulation (Edvinsson et al. 1986). Instead CGRP was demonstrated to play a key role in a protective trigeminovascular reflex (McCulloch et al. 1986), whereby CGRP is released by trigeminal nerves in response to local cerebral vasoconstriction to cause dilation and maintain cerebral blood flow. These findings suggested that CGRP was involved in migraine pathophysiology (Edvinsson 1985). The decisive results for demonstration of its involvement in primary headaches came a few years later: CGRP is selectively released from the trigeminal system during acute migraine headache attacks, and this release is prevented by anti-migraine drugs such as triptans (Edvinsson et al. 2018).

3   Components of CGRP Transmission Relevant to Migraine Therapies

In CGRP nerves, the presynaptic terminals take the form of focal swellings, called axonal varicosities that occur at regular intervals along the axon like a strand of pearls. Upon nerve stimulation or depolarization, CGRP is released from its storage vesicles via calcium-dependent exocytosis. Capsaicin, a component of chili peppers, also will cause release of α-CGRP, and this compound has been useful as an experimental tool to release and ultimately deplete the peptide from CGRP nerves. Presynaptic receptors located on trigeminal neurons regulate CGRP release. Presynaptic serotonin 5-HT_{1B} and 5-HT_{1D} receptors (Hou et al. 2001) may inhibit CGRP release, and they are of particular relevance to migraine treatment. These receptors are the target for the therapeutic effects of the triptan drugs, e.g., sumatriptan, in relief of migraine headache, while we consider the vasomotor response to triptans mainly as an unnecessary bi-product (Edvinsson et al. 2018).

In bipolar trigeminal sensory neurons located in the trigeminal ganglion, CGRP is released at both peripheral and central nerve terminals. In the trigeminovascular system, the peripheral branch of CGRP axons form perivascular nerves that run along the adventitial-medial border in the wall of intracranial cerebral and dural blood vessels (Eftekhari et al. 2013). Axon varicosities are separated from the adjacent smooth muscle cell by a relatively wide cleft (100–500 nm). CGRP fibers also terminate in nonvascular regions of the dura mater, where they play a role in meningeal nociception.

The central branch of trigeminovascular CGRP axons project to the spinal trigeminal nucleus and C1–2-levels of the spinal cord, notably laminae I and II of the dorsal horn (Edvinsson 2011; Eftekhari and Edvinsson 2011). Central CGRP release sites, as demonstrated by immunohistochemical co-localization with the synaptic vesicle protein synaptophysin (Eftekhari and Edvinsson 2011), also exhibit a beaded appearance consistent with axonal varicosities. In addition to the peripheral and central terminations, GGRP is released within the trigeminal ganglion itself, where it likely regulates sensory processing at an early stage in the pain pathway. Within the ganglion, CGRP is expressed in thin, beaded fibers that are likely release sites on local branches of CGRP axons (Edvinsson et al. 2018).

Following release, CGRP is broken down by metalloproteases (Kim et al. 2013). Amidation at the C-terminus helps protect the peptide and increases its half-life. Human plasma levels are usually within the low picomolar range and generally attributed to spillover from sites of neuronal release. The plasma half-life of CGRP in humans was estimated following CGRP infusion to be 7 min for a fast decay phase and 26 min for a slower phase of decay (Kraenzlin et al. 1985). Release of CGRP after stimulation of the trigeminal ganglion in humans has been measured in blood collected from the external jugular vein draining the extracerebral tissues (Goadsby et al. 1988). Because of the instability of the peptide, the blood samples were collected close to the release site using a protocol to limit proteolysis.

4   CGRP Receptor Family

In line with other G protein-coupled receptors the CGRP receptor can trigger diverse signaling pathways and undergo regulatory control (Hay et al. 2018). The CGRP receptor consists of a seven transmembrane G protein-coupled calcitonin receptor-like receptor (CLR) with a single membrane-spanning receptor activity modifying protein (RAMP1) (Hay et al. 2018). CLR is a member of the secretin receptor family (Class B GPCR), and it is a required element in receptors for CGRP and adrenomedullin (AM_1 and AM_2). In order to create a functional membrane receptor with specific affinity for CGRP, CLR must form a heterodimer with the RAMP1 (McLatchie et al. 1998). RAMP proteins alter the pharmacology, functionality, and cell trafficking of specific GPCRs. The ligand-binding domain of the CGRP receptor is located at the interface between the RAMP1 and CLR proteins (Hay et al. 2018; Mallee et al. 2002). Thus, co-expression of both CLR and RAMP1 is necessary for a cell to respond to CGRP. Recently Liang and colleagues reported the detailed structure of the human CGRP receptor (Liang et al. 2018). The work illustrated the site of binding of CGRP and gepants at the interface of CLR and RAMP1. If CLR is coupled with RAMP2 or RAMP3, the two forms of adrenomedullin receptors, AM_1 and AM_2, are formed, respectively.

The other part of this group of receptors is that based on calcitonin. CT per se can act as agonist at the CT receptor (CTR). The CTR does not require combination with a RAMP molecule. When the CTR is combined with either of the RAMPs the amylin family of receptors is formed: AMY_1 (RAMP1), AMY_2 (RAMP2), and AMY_3 (RAMP3) (Hay et al. 2018). The role of the different receptors and ligands of the CGRP family is less well understood.

The CGRP receptor complex includes two cytoplasmic proteins that associate with the CLR-RAMP1 heterodimer to mediate signal transduction. CLR is coupled to a G-protein containing the Gs α subunit (Gαs) that activates adenylyl cyclase and cAMP-dependent signaling pathways. In addition, the CGRP receptor associates with a receptor coupling protein (RCP) that amplifies G-protein activation and is important for optimal signal transduction (Egea and Dickerson 2012). Receptor-mediated increases in intracellular cAMP activate protein kinase A (PKA), resulting in the phosphorylation of multiple downstream targets, including potassium-sensitive ATP channels (K_{ATP} channels), extracellular signal-related kinases (ERKs), or transcription factors such as cAMP response element-binding protein (CREB). In cerebrovascular smooth muscle, elevation of cAMP by CGRP results in vasorelaxation.

An important feature of CGRP signaling is the regulation and desensitization of the receptor following agonist activation. After CGRP has bound to its receptor, the CLR component is rapidly phosphorylated; and the receptor is subsequently internalized via recruitment of β-arrestin. Classically, the receptor binding is thought to signal only at the cell surface but is now recognized that some G protein-coupled receptors including the CGRP receptor undergo sustained signaling from endosomes, once internalizes in response to the ligand to signal pain (Yarwood et al. 2017). Studies using labeled receptor components were found to co-localize together suggesting that both receptor elements co-internalize (Kuwasako et al. 2000; Padilla et al. 2007). Thus, transient stimulation by CGRP induces internalization of the receptor to endosomes that allows rapid recycling back to the plasma membrane. Chronic exposure to CGRP, however, initiates an internalization process that traffics the receptor to lysosomes for degradation. Thus, the CGRP receptor can signal not only via the cell membrane but also from endosomes within cells. The regulation of CGRP receptors in disease is another challenge for researchers.

5 CGRP Receptor Antagonists

Migraine is a prevalent, disabling neurological disorder involving the trigeminovascular system. Previous and current treatments were either originally intended for other conditions and/or associated with intolerable adverse effects. As stated above CGRP is the most prevalent neuropeptide in the trigeminal system and plays an important role in the pathophysiology of migraine (Edvinsson et al. 2018). The gepants and the monoclonal antibodies are the first treatments created specifically for migraine, modulating pain signaling pathways.

The first specific antagonist for CGRP receptors was olcegepant which was soon found to be very potent and specific, but was a di-peptide and hence only works when given parenterally (Doods et al. 2000). A number of other small molecule non-peptides antagonists have since been identified that potently and selectively block CGRP responses in both experimental and clinical studies (Holland and Goadsby 2018). Although chemically unrelated, this class of antagonists is collectively called "gepants" due to their common mode of action. Significantly, all gepants tested to date are efficacious in migraine patients. Interestingly, a common feature of the clinically relevant gepants is their high affinity for CGRP receptors of humans and nonhuman primates relative to other species. This is due to a species-specific residue located at the interface between the RAMP1 and CLR proteins, indicating this region as the site of antagonist binding (Mallee et al. 2002). The recent characterization of the CGRP receptor structure has revealed the exact location of the binding of CGRP and gepants to the receptor complex (Liang et al. 2018).

Gepants potently block the binding of CGRP to its receptor in cells and in tissue preparations and produced a rightward shift of CGRP concentration-response curves in signaling (cAMP production) and functionality (vasodilation). The newer gepants that has been developed still exhibited high potency and good oral bioavailability, e.g., telcagepant (Paone et al. 2007), but were discontinued due to liver enzyme issues (Edvinsson and Linde 2010). Currently, three of the newer gepants, ubrogepant, aterogepant, and rimegepant, have passed phase III testing and are now in the finale for registration. It will be interesting to learn how they will be positioned in the therapy of migraine.

6   CGRP and CGRP Receptor Antibodies for Prophylaxis

An alternative strategy for blocking CGRP transmission in migraine patients is to use selective monoclonal antibodies that bind either CGRP or the CGRP receptor. This approach has been remarkably successful for decreasing the frequency of migraine attacks. Four such antibodies have now completed Phase III clinical trials and are in the process of approval worldwide.

The therapeutic goal for migraine prophylaxis is to reduce the number of migraine days experienced by patients who suffer frequent attacks. Antibodies to either CGRP or the CGRP receptor have demonstrated efficacy in reducing migraine days in patients with episodic (<15 days/month) or chronic (>15 days/month) migraine. The anti-receptor antibody Erenumab (Aimovig) is a human IgG2 monoclonal antibody targeted to the CGRP binding site on the CGRP receptor. It is administered once a month by subcutaneous injection to effectively reduce migraine frequency. Erenumab is approved in the USA by FDA (May 2018) and by EMA for Europe, and is on the market. Currently, there are three different monoclonal antibodies targeted to sites on the CGRP peptide itself, which have all concluded Phase III testing for migraine prevention. Overall, these anti-CGRP antibodies appear to show similar efficacy and safety profiles. Eptinezumab is a genetically engineered humanized IgG1 monoclonal antibody that is formulated for intravenous administration and intended for once per quarter dosing. The other two antibodies to CGRP have now been approved by the FDA (September 2018): Fremanezumab (Ajovy) and Galcanezumab (Emgality). Fremanezumab is a fully humanized IgG2a monoclonal antibody that is given once per month by subcutaneous injection. Galcanezumab is a fully humanized IgG4 monoclonal antibody that also is administered subcutaneously on a monthly basis to reduce the number of migraine days.

The antibodies are particularly suited for use as a prophylactic treatment for migraine with the advantages of patient adherence and tolerability. They have a prolonged serum half-life (20–30 days) that enables patients to take their medication less frequently for prevention of migraine attacks. Antibodies bind their target site with high affinity and selectivity, thus reducing the potential for unwanted, off-target effects. In contrast to small exogenous molecules such as the gepants, antibodies are not processed by the liver, thus avoiding the potential for liver toxicities and hepatic drug interactions. No adverse cardiovascular or cerebrovascular effects have been reported for these antibodies. A primary disadvantage of the antibodies, however, is they are not orally active and must be administered by injection. Injection-site reactions, including pain, are the most commonly reported adverse events. These reactions are usually mild and transient, and are less likely with intravenous as compared to subcutaneous administration. Thus, overall, the anti-migraine antibodies have been shown in clinical trials to be well tolerated and safe in patients. Because no head-to-head comparisons have been done, it is difficult to discuss if one or the other is to prefer; however, published data show more or less similar results.

7   Conclusion

The development of the understanding of the CGRP family of peptides and receptors has shed new light on migraine pathophysiology and provided several options to treat this large group of patients with migraine. The promising new gepants and antibody medications are on their way to the patients and it will be rewarding to observe their clinical effects in the general population.

References

Amara SG, Jonas V, Rosenfeld MG, Ong ES, Evans RM (1982) Alternative RNA processing in calcitonin gene expression generates mRNAs encoding different polypeptide products. Nature 298:240–244

Charles A (2018) The pathophysiology of migraine: implications for clinical management. Lancet Neurol 17:174–182

Dodick DW (2018) Migraine. Lancet 391:1315–1330

Doods H, Hallermayer G, Wu D, Entzeroth M, Rudolf K, Engel W, Eberlein W (2000) Pharmacological profile of BIBN4096BS, the first selective small molecule CGRP antagonist. Br J Pharmacol 129:420–423

Edvinsson L (1985) Functional-role of perivascular peptides in the control of cerebral-circulation. Trends Neurosci 8:126–131

Edvinsson L (2011) Tracing neural connections to pain pathways with relevance to primary headaches. Cephalalgia 31:737–747

Edvinsson L, Linde M (2010) New drugs in migraine treatment and prophylaxis: telcagepant and topiramate. Lancet 376:645–655

Edvinsson L, Uddman R (2005) Neurobiology in primary headaches. Brain Res Brain Res Rev 48:438–456

Edvinsson LM, McCulloch J, Kingman TA, Uddman R (1986) On the functional role of the trigemino-cerebrovascular system in the regulation of the cerebral circulation. In: Owman CH, Hardebo JE (eds) Neural regulation of brain circulation. Elsevier, Amsterdam, pp 407–418

Edvinsson L, Hara H, Uddman R (1989) Retrograde tracing of nerve fibers to the rat middle cerebral artery with true blue: colocalization with different peptides. J Cereb Blood Flow Metab 9:212–218

Edvinsson L, Haanes KA, Warfvinge K, Krause DN (2018) CGRP as the target of new migraine therapies - successful translation from bench to clinic. Nat Rev Neurol 14:338–350

Eftekhari S, Edvinsson L (2011) Calcitonin gene-related peptide (CGRP) and its receptor components in human and rat spinal trigeminal nucleus and spinal cord at C1-level. BMC Neurosci 12:112

Eftekhari S, Salvatore CA, Calamari A, Kane SA, Tajti J, Edvinsson L (2010) Differential distribution of calcitonin gene-related peptide and its receptor components in the human trigeminal ganglion. Neuroscience 169:683–696

Eftekhari S, Warfvinge K, Blixt FW, Edvinsson L (2013) Differentiation of nerve fibers storing CGRP and CGRP receptors in the peripheral trigeminovascular system. J Pain 14:1289–1303

Egea SC, Dickerson IM (2012) Direct interactions between calcitonin-like receptor (CLR) and CGRP-receptor component protein (RCP) regulate CGRP receptor signaling. Endocrinology 153:1850–1860

Goadsby PJ, Edvinsson L (1993) The trigeminovascular system and migraine: studies characterizing cerebrovascular and neuropeptide changes seen in humans and cats. Ann Neurol 33:48–56

Goadsby PJ, Edvinsson L, Ekman R (1988) Release of vasoactive peptides in the extracerebral circulation of humans and the cat during activation of the trigeminovascular system. Ann Neurol 23:193–196

Goadsby PJ, Edvinsson L, Ekman R (1990) Vasoactive peptide release in the extracerebral circulation of humans during migraine headache. Ann Neurol 28:183–187

Goadsby PJ, Holland PR, Martins-Oliveira M, Hoffmann J, Schankin C, Akerman S (2017) Pathophysiology of migraine: a disorder of sensory processing. Physiol Rev 97:553–622

Hara H, Jansen I, Ekman R, Hamel E, MacKenzie ET, Uddman R, Edvinsson L (1989) Acetylcholine and vasoactive intestinal peptide in cerebral blood vessels: effect of extirpation of the sphenopalatine ganglion. J Cereb Blood Flow Metab 9:204–211

Hay DL, Garelja ML, Poyner DR, Walker CS (2018) Update on the pharmacology of calcitonin/CGRP family of peptides: IUPHAR Review 25. Br J Pharmacol 175:3–17

Holland PR, Goadsby PJ (2018) Targeted CGRP small molecule antagonists for acute migraine therapy. Neurotherapeutics 15:304–312

Hou M, Kanje M, Longmore J, Tajti J, Uddman R, Edvinsson L (2001) 5-HT(1B) and 5-HT(1D) receptors in the human trigeminal ganglion: co-localization with calcitonin gene-related peptide, substance P and nitric oxide synthase. Brain Res 909:112–120

Kim YG, Lone AM, Saghatelian A (2013) Analysis of the proteolysis of bioactive peptides using a peptidomics approach. Nat Protoc 8:1730–1742

Kraenzlin ME, Ch'ng JL, Mulderry PK, Ghatei MA, Bloom SR (1985) Infusion of a novel peptide, calcitonin gene-related peptide (CGRP) in man. Pharmacokinetics and effects on gastric acid secretion and on gastrointestinal hormones. Regul Pept 10:189–197

Kuwasako K, Shimekake Y, Masuda M, Nakahara K, Yoshida T, Kitaura M, Kitamura K, Eto T, Sakata T (2000) Visualization of the calcitonin receptor-like receptor and its receptor activity-modifying proteins during internalization and recycling. J Biol Chem 275:29602–29609

Liang YL, Khoshouei M, Deganutti G, Glukhova A, Koole C, Peat TS, Radjainia M, Plitzko JM, Baumeister W, Miller LJ et al (2018) Cryo-EM structure of the active, Gs-protein complexed, human CGRP receptor. Nature 561:492–497

Mallee JJ, Salvatore CA, LeBourdelles B, Oliver KR, Longmore J, Koblan KS, Kane SA (2002) Receptor activity-modifying protein 1 determines the species selectivity of non-peptide CGRP receptor antagonists. J Biol Chem 277:14294–14298

May A (2017) Understanding migraine as a cycling brain syndrome: reviewing the evidence from functional imaging. Neurol Sci 38:125–130

May A, Schulte LH (2016) Chronic migraine: risk factors, mechanisms and treatment. Nat Rev Neurol 12:455–464

McCulloch J, Uddman R, Kingman TA, Edvinsson L (1986) Calcitonin gene-related peptide: functional role in cerebrovascular regulation. Proc Natl Acad Sci U S A 83:5731–5735

McLatchie LM, Fraser NJ, Main MJ, Wise A, Brown J, Thompson N, Solari R, Lee MG, Foord SM (1998) RAMPs regulate the transport and ligand specificity of the calcitonin-receptor-like receptor. Nature 393:333–339

Padilla BE, Cottrell GS, Roosterman D, Pikios S, Muller L, Steinhoff M, Bunnett NW (2007) Endothelin-converting enzyme-1 regulates endosomal sorting of calcitonin receptor-like receptor and beta-arrestins. J Cell Biol 179:981–997

Paone DV, Shaw AW, Nguyen DN, Burgey CS, Deng JZ, Kane SA, Koblan KS, Salvatore CA, Mosser SD, Johnston VK et al (2007) Potent, orally bioavailable calcitonin gene-related peptide receptor antagonists for the treatment of migraine: discovery of N-[(3R,6S)-6-(2,3-difluorophenyl)-2-oxo-1- (2,2,2-trifluoroethyl)azepan-3-yl]-4- (2-oxo-2,3-dihydro-1H-imidazo[4,5-b]pyridin-1-yl)piperidine-1-carboxamide (MK-0974). J Med Chem 50:5564–5567

Russell FA, King R, Smillie SJ, Kodji X, Brain SD (2014) Calcitonin gene-related peptide: physiology and pathophysiology. Physiol Rev 94:1099–1142

Tajti J, Uddman R, Moller S, Sundler F, Edvinsson L (1999) Messenger molecules and receptor mRNA in the human trigeminal ganglion. J Auton Nerv Syst 76:176–183

Uddman R, Edvinsson L (1989) Neuropeptides in the cerebral circulation. Cerebrovasc Brain Metab Rev 1:230–252

Uddman R, Edvinsson L, Ekman R, Kingman T, McCulloch J (1985) Innervation of the feline cerebral vasculature by nerve fibers containing calcitonin gene-related peptide: trigeminal origin and co-existence with substance P. Neurosci Lett 62:131–136

Uddman R, Edvinsson L, Ekblad E, Hakanson R, Sundler F (1986a) Calcitonin gene-related peptide (CGRP): perivascular distribution and vasodilatory effects. Regul Pept 15:1–23

Uddman R, Edvinsson L, Jansen I, Stiernholm P, Jensen K, Olesen J, Sundler F (1986b) Peptide-containing nerve fibres in human extracranial tissue: a morphological basis for neuropeptide involvement in extracranial pain? Pain 27:391–399

Weiller C, May A, Limmroth V, Juptner M, Kaube H, Schayck RV, Coenen HH, Diener HC (1995) Brain stem activation in spontaneous human migraine attacks. Nat Med 1:658–660

Yarwood RE, Imlach WL, Lieu T, Veldhuis NA, Jensen DD, Klein Herenbrink C, Aurelio L, Cai Z, Christie MJ, Poole DP, Porter CJH, McLean P, Hicks GA, Geppetti P, Halls ML, Canals M, Bunnett NW (2017) Endosomal signaling of the receptor for calcitonin gene-related peptide mediates pain transmission. Proc Natl Acad Sci U S A 114:12309–12314

Understanding CGRP and Cardiovascular Risk

Eloísa Rubio-Beltrán and Antoinette Maassen van den Brink

Contents

Abstract

Increasing knowledge about the role of calcitonin gene-related peptide (CGRP) in migraine pathophysiology has led to the development of antibodies against this peptide or its receptor. However, CGRP is widely expressed throughout the body, participating not only in pathophysiological conditions but also in several physiological processes and homeostatic responses during pathophysiological events. Therefore, in this chapter, the risks of long-term blockade of the CGRP pathway will be discussed, with focus on the cardiovascular system, as this peptide has been described to have a protective role during ischemic events, and migraine patients present a higher risk of stroke and myocardial infarction.

Keywords

Cardiovascular safety · CGRP · CGRP (receptor) antibodies · Migraine · Myocardial infarction · Stroke

E. Rubio-Beltrán · A. M. van den Brink (✉)
Division of Vascular Medicine and Pharmacology, Department of Internal Medicine, Erasmus Medical Center, Rotterdam, The Netherlands
e-mail: a.vanharen-maassenvandenbrink@erasmusmc.nl

© Springer Nature Switzerland AG 2019
S. D. Brain, P. Geppetti (eds.), *Calcitonin Gene-Related Peptide (CGRP) Mechanisms*,
Handbook of Experimental Pharmacology 255, https://doi.org/10.1007/164_2019_204

1 Introduction

Migraine is a highly disabling neurovascular disorder (Stovner et al. 2018). As mentioned in the previous chapters, calcitonin gene-related peptide (CGRP) has been described to play an important role in migraine pathophysiology (Edvinsson 2017; Goadsby et al. 2002). As a result, the CGRP pathway has become a promising target.

Initially, CGRP receptor antagonists (gepants) were developed for the acute treatment of migraine and proved to be effective (Doods et al. 2000; Edvinsson and Linde 2010). Unfortunately, pharmacokinetic limitations and hepatotoxicity cases did not allow the initial gepants to reach the market (Negro et al. 2012). New gepants are currently in Phase II trials for the acute and prophylactic treatment of migraine, with no hepatotoxicity reported (Holland and Goadsby 2018; Tepper 2018); nevertheless, the concerns about the hepatotoxicity reports led to the development of CGRP (receptor) antibodies for the prophylactic treatment of migraine (Deen et al. 2017; Schuster et al. 2015; Wrobel Goldberg and Silberstein 2015). Preliminary results of the clinical trials are promising and have not reported serious side effects (Mitsikostas and Reuter 2017); however, it is important to consider the physiological role of this peptide and the possible side effects after long-term blockade of the CGRP pathway.

In this chapter, the role of CGRP in physiological processes will be described, with focus on the cardiovascular system, as migraine patients present a higher risk of stroke and myocardial infarction (Etminan et al. 2005; Kurth et al. 2009; Sacco et al. 2013; Scher et al. 2005).

2 CGRP and the Cardiovascular System

CGRP and its fibers are widely distributed in peripheral and central structures. In the cardiovascular system, sensory CGRPergic fibers have been described to innervate the blood vessels and the heart (Opgaard et al. 1995; Uddman et al. 1986; Wimalawansa and MacIntyre 1988). Several studies have shown that CGRP plays an important role in the regulation of blood pressure and in the homeostatic responses during ischemic events and hypertension (HT) (Edvinsson et al. 1998; Keith et al. 2000; Lindstedt et al. 2006; MaassenVanDenBrink et al. 2016; McCulloch et al. 1986; Russell et al. 2014).

2.1 CGRP and Hypertension

As mentioned above, CGRP has been demonstrated to be involved in the regulation of blood pressure. Although its role under physiological conditions may be limited (Smillie and Brain 2011), it seems to act as a protective/compensatory mechanism during HT (Smillie et al. 2014). In accordance with this hypothesis, in the deoxycorticosterone-salt HT model, CGRP knockout mice had a significant increase in 24-h mean arterial pressure (MAP) and renal damage when compared to wild types (Jianping et al. 2013), while in non-treated animals, only the 7-day average of the daytime MAP was significantly increased (Mai et al. 2014). Moreover, in a model of angiotensin II-induced HT, CGRP knockout mice exhibited an enhanced increase in MAP and aortic hypertrophy. This was accompanied by an upregulation of the CGRP receptor components expression, reinforcing the role of CGRP release as a safeguard mechanism against the onset and maintenance of HT (Smillie et al. 2014). This increase in blood pressure has been associated to an elevated sympathetic activation, as CGRP knockout mice show an increase in urine and plasma markers of catecholamine release (Mai et al. 2014). Indeed, bolus injections of the CGRP antagonist olcegepant enhance the vasopressor sympathetic outflow in pithed rats (Avilés-Rosas et al. 2017). Moreover, CGRP is not only involved in peripheral mechanisms, but it also participates in the maintenance of cerebrovascular reactivity during chronic HT (Wang et al. 2015).

The abovementioned studies support the role of CGRP in blood pressure regulation during HT. As a result, a novel CGRP analogue was recently developed to improve and reverse cardiovascular disease. Results from in vivo preclinical models of hypertension and cardiac failure showed positive antihypertensive effects, an attenuation of cardiac remodeling, and an increase in angiogenesis and cell survival after administration of the CGRP analogue (Aubdool et al. 2017).

2.2 CGRP and Ischemia

During severe HT and focal cerebral ischemia, CGRP has been demonstrated to act as a neuroprotector, by increasing cerebral blood flow (Moskowitz et al. 1989; Sakas et al. 1989; Zhang et al. 2011). In rats, if CGRP is administrated at the beginning of reperfusion after experimental cerebral artery occlusion, a reduction in brain edema is observed, probably due to a decrease in the blood-brain barrier disruption (Liu et al. 2011). In patients with subarachnoid hemorrhage (SAH), higher levels of plasma CGRP have been associated with delayed vasospasm (Juul et al. 1990) and infusion of CGRP further reduced vasospasm (Juul et al. 1994). Similarly, in another cohort of patients with SAH, CGRP levels in cerebrospinal fluid of patients without vasospasm were significantly higher than the levels of patients with vasospasm, with the former group not developing cerebral ischemia (Schebesch et al. 2013). In an experimental rat model of SAH, CGRP expression was decreased; however, an enhanced CGRP-dependent vasodilation was observed (Edvinsson et al. 1990). Finally, vasospasm after induction of SAH by placing a clot around the internal carotid artery bifurcation was significantly ameliorated in monkeys that were treated with slow-release CGRP tablets, consisting of compressed microspheres containing CGRP, and that were placed in the cerebrospinal fluid (Inoue et al. 1996). Due to their composition, these compressed microsphere tablets released CGRP for a period of several weeks, providing proof-of-concept data suggesting CGRP agonism as a possible therapeutic target for SAH patients.

In myocardial ischemia, CGRP is also considered to be released as a protective mechanism. Preclinical studies in rats and mice show protective hemodynamic and metabolic changes mediated by CGRP in response to ischemic events (Chai et al. 2006; Gao et al. 2015; Homma et al. 2014; Lei et al. 2016). Moreover, in clinical studies, intravenous administration of CGRP resulted in a decrease of both systolic and diastolic arterial pressure and an increase of heart rate (Gennari and Fischer 1985). Furthermore, when infused in patients with congestive heart failure, myocardial contractility is improved (Gennari et al. 1990). Interestingly, lower plasma levels of CGRP have been reported in patients with diabetes mellitus and coronary artery disease, when compared to controls, suggesting an alteration in the CGRP (cardioprotective) pathway (Wang et al. 2012). Obviously, these observations need to be confirmed in future, and it should be elucidated whether potential changes in patients with cardiovascular disease reflect a cause or consequence of this disease.

2.3   CGRP and Preeclampsia

CGRP also seems to be involved in the vascular adaptations during pregnancy, as plasma levels increase through the gestation period, reaching their maximum during the last trimester and normalizing after delivery. However, in preeclampsia, a pregnancy disorder characterized by high blood pressure and proteinuria, CGRP levels are lower (Yadav et al. 2014). The mechanisms behind this are not yet known but indicate an alteration in the CGRP signaling, similar as observed in patients with cardiovascular disease.

3   Cardiovascular Risk and Migraine

Numerous studies have shown that migraine patients present an increased risk of hemorrhagic and ischemic stroke, with the risk being higher for women (Chang et al. 1999; Etminan et al. 2005; Sacco et al. 2013; Schurks et al. 2009; Spector et al. 2010; Tzourio et al. 1995). Moreover, a higher risk of myocardial infarction, coronary artery disease, and altered arterial function has also been described (Scher et al. 2005; Vanmolkot et al. 2007). Unfortunately, the mechanisms behind these increases are not clear, but it is thought to involve genetic aspects and vascular dysfunction, among other factors. This poses a concern, as currently the main novel therapeutic target for migraine treatment is blocking CGRP or its receptor, which could increase cardiovascular risk (Deen et al. 2017; MaassenVanDenBrink et al. 2016).

3.1   Cardiovascular Risk, Migraine, and Women

Migraine is almost three times more prevalent in women than in men (Buse et al. 2013). Frequency, intensity of headaches, disability, and chronification have also

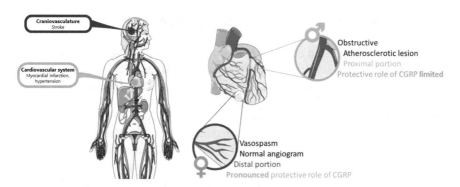

Fig. 1 Theoretical concerns after long-term blockade of CGRP (receptor). Migraine patients present an increased risk of cardiovascular disease, and CGRP participates as a safeguard during ischemic events which suggest that after CGRP blockade, the (cardio)vascular risk could increase further. In myocardial ischemia, CGRP seems to have a more prominent role in the distal portion than in the proximal portion of the coronary arteries, which may represent a downside for women, as ischemic events in the distal portion are more common in female patients, while proximal obstructions are more prevalent in male patients

been reported to be higher in female patients (Buse et al. 2013; Labastida-Ramirez et al. 2017). In addition, women with migraine present a higher risk of stroke when compared to men with migraine, and, as before menopause the prevalence of cardiovascular events is rather low, after menopause the occurrence rises sharply (Bushnell et al. 2014; Mieres et al. 2014).

In myocardial infarction, sex-related differences have also been observed. Women usually present angina-like chest pain and a positive response to stress testing but no visible obstructions during angiography as it is caused by vasospasms of the small intramyocardial portions of the coronary arteries (Humphries et al. 2008; Kaski et al. 1995). On the contrary, men usually present with occlusions of the proximal conducting portion, which are evident during an angiography (Fig. 1). This disparity may represent a downside for female migraine patients undergoing treatment with CGRP (receptor) blockade, as CGRP-dependent vasodilation (and cardioprotection) in coronary arteries is more pronounced in the distal portions than in the proximal portions (Chan et al. 2010; Gulbenkian et al. 1993; MaassenVanDenBrink et al. 2016). Moreover, CGRP signaling seems to be modulated by ovarian steroid hormones, as women have higher plasma levels than men, and the levels increase when patients are under contraceptives (Valdemarsson et al. 1990). Furthermore, the decrease in blood pressure and the positive inotropic effect induced by CGRP administration have been described to be enhanced when 17β-estradiol or progesterone is co-administered (Al-Rubaiee et al. 2013; Gangula et al. 2002). This evidence, taken together, strongly suggests a (protective) synergistic interaction between ovarian steroid hormones and CGRP and reiterates the concerns about CGRP (receptor) blockade in women, as this could increase their risk of suffering an ischemic event even more, especially after menopause.

4   Safety Assessment of CGRP Blockade

Considering the increased cardiovascular risk of migraine patients discussed in the previous section, it is important to perform studies that *correctly* assess the safety of CGRP (receptor) blockade. For such a purpose, cardiovascularly compromised subjects should be included that properly represent the population of migraine patients potentially using these drugs.

Unfortunately, even though the grand majority of the CGRP (receptor) antibodies have been approved, currently only one group has evaluated their cardiovascular safety profile in cardiovascularly compromised patients (Depre et al. 2018). In this study, a randomized, double-blind, placebo-controlled trial was performed to evaluate the effect of erenumab, a fully human monoclonal antibody directed against the CGRP receptor, on exercise time during a treadmill test in patients with stable angina pectoris. The authors reported no alterations in performance between patients receiving erenumab and placebo. Apart from serious pharmacological concerns about the validity of this specific study, because no evidence was presented on whether effective CGRP receptor blockade was achieved at the time of the treadmill test (Maassen van den Brink et al. 2018), the study population needs further attention.

In the study from Depre et al., the patients included suffered from stable angina pectoris, most likely due to stenosis of the epicardial conducting portions of the coronary artery. As discussed previously, the role of CGRP is limited in the proximal coronary artery (Chan et al. 2010). Whereas most patients using the antibodies will be female, this study included 78% males, as stable angina related to epicardial stenosis is mainly present in male patients. Thus, women, who pose a major concern and may suffer from microvascular disease, where CGRP may be a relevant mediator, were underrepresented in this study.

While in some cases performing appropriate studies in relevant patient groups may be ethically and practically challenging, preclinical studies are excellent to shed more light on the role of CGRP in cardiovascular regulation. In this light, it is important also to take into account potential differences between short-term and long-term blockade of CGRP or its receptor in models of cardiovascular disease in both male and female animals.

5   Conclusion

CGRP plays an important role in (cardio)vascular protection. However, it is also involved in migraine pathophysiology, and the current novel treatments involve CGRP (receptor) blockade. As migraine patients present higher cardiovascular risk, with women at higher risk, chronic blockade of the CGRP pathway poses a concern. While the initial clinical trials don't indicate frequent adverse events, it is of crucial importance to correctly evaluate the safety profile of these novel drugs, in order to prevent serious adverse effects when these drugs will be used on a large scale.

References

Al-Rubaiee M, Gangula PR, Millis RM, Walker RK, Umoh NA, Cousins VM, Jeffress MA, Haddad GE (2013) Inotropic and lusitropic effects of calcitonin gene-related peptide in the heart. Am J Physiol Heart Circ Physiol 304:H1525–H1537

Aubdool AA, Thakore P, Argunhan F, Smillie SJ, Schnelle M, Srivastava S, Alawi KM, Wilde E, Mitchell J, Farrell-Dillon K, Richards DA, Maltese G, Siow RC, Nandi M, Clark JE, Shah AM, Sams A, Brain SD (2017) A novel alphacalcitonin gene-related peptide analogue protects against end-organ damage in experimental hypertension, cardiac hypertrophy, and heart failure. Circulation 136:367–383

Avilés-Rosas VH, Rivera-Mancilla E, Marichal-Cancino BA, Manrique-Maldonado G, Altamirano-Espinoza AH, Maassen Van Den Brink A, Villalón CM (2017) Olcegepant blocks neurogenic and non-neurogenic CGRPergic vasodepressor responses and facilitates noradrenergic vasopressor responses in pithed rats. Br J Pharmacol 174:2001–2014

Buse DC, Loder EW, Gorman JA, Stewart WF, Reed ML, Fanning KM, Serrano D, Lipton RB (2013) Sex differences in the prevalence, symptoms, and associated features of migraine, probable migraine and other severe headache: results of the American Migraine Prevalence and Prevention (AMPP) Study. Headache 53:1278–1299

Bushnell C, McCullough LD, Awad IA, Chireau MV, Fedder WN, Furie KL, Howard VJ, Lichtman JH, Lisabeth LD, Piña IL, Reeves MJ, Rexrode KM, Saposnik G, Singh V, Towfighi A, Vaccarino V, Walters MR (2014) Guidelines for the prevention of stroke in women. Stroke 45:1545–1588

Chai W, Mehrotra S, Jan Danser AH, Schoemaker RG (2006) The role of calcitonin gene-related peptide (CGRP) in ischemic preconditioning in isolated rat hearts. Eur J Pharmacol 531:246–253

Chan KY, Edvinsson L, Eftekhari S, Kimblad PO, Kane SA, Lynch J, Hargreaves RJ, de Vries R, Garrelds IM, van den Bogaerdt AJ, Danser AH, Maassenvandenbrink A (2010) Characterization of the calcitonin gene-related peptide receptor antagonist telcagepant (MK-0974) in human isolated coronary arteries. J Pharmacol Exp Ther 334:746–752

Chang CL, Donaghy M, Poulter N (1999) Migraine and stroke in young women: case-control study. Br Med J 318:13–18

Deen M, Correnti E, Kamm K, Kelderman T, Papetti L, Rubio-Beltran E, Vigneri S, Edvinsson L, Maassen Van Den Brink A, European Headache Federation School of Advanced Studies (2017) Blocking CGRP in migraine patients – a review of pros and cons. J Headache Pain 18:96

Depre C, Antalik L, Starling A, Koren M, Eisele O, Lenz RA, Mikol DD (2018) A randomized, double-blind, placebo-controlled study to evaluate the effect of erenumab on exercise time during a treadmill test in patients with stable angina. Headache 58:715–723

Doods H, Hallermayer G, Wu D, Entzeroth M, Rudolf K, Engel W, Eberlein W (2000) Pharmacological profile of BIBN4096BS, the first selective small molecule CGRP antagonist. Br J Pharmacol 129:420–423

Edvinsson L (2017) The trigeminovascular pathway: role of CGRP and CGRP receptors in migraine. Headache 57(Suppl 2):47–55

Edvinsson L, Linde M (2010) New drugs in migraine treatment and prophylaxis: telcagepant and topiramate. Lancet 376:645–655

Edvinsson L, Delgado-Zygmunt T, Ekman R, Jansen I, Svendgaard NA, Uddman R (1990) Involvement of perivascular sensory fibers in the pathophysiology of cerebral vasospasm following subarachnoid hemorrhage. J Cereb Blood Flow Metab 10:602–607

Edvinsson L, Mulder H, Goadsby PJ, Uddman R (1998) Calcitonin gene-related peptide and nitric oxide in the trigeminal ganglion: cerebral vasodilatation from trigeminal nerve stimulation involves mainly calcitonin gene-related peptide. J Auton Nerv Syst 70:15–22

Etminan M, Takkouche B, Isorna FC, Samii A (2005) Risk of ischaemic stroke in people with migraine: systematic review and meta-analysis of observational studies. Br Med J 330:63

Gangula PRR, Wimalawansa SJ, Yallampalli C (2002) Sex steroid hormones enhance hypotensive effects of calcitonin gene-related peptide in aged female rats. Biol Reprod 67:1881–1887

Gao Y, Song J, Chen H, Cao C, Lee C (2015) TRPV1 activation is involved in the cardioprotection of remote limb ischemic postconditioning in ischemia-reperfusion injury rats. Biochem Biophys Res Commun 463:1034–1039

Gennari C, Fischer JA (1985) Cardiovascular action of calcitonin gene-related peptide in humans. Calcif Tissue Int 37:581–584

Gennari C, Nami R, Agnusdei D, Fischer JA (1990) Improved cardiac performance with human calcitonin gene related peptide in patients with congestive heart failure. Cardiovasc Res 24:239–241

Goadsby PJ, Lipton RB, Ferrari MD (2002) Migraine – current understanding and treatment. N Engl J Med 346:257–270

Gulbenkian S, Opgaard OS, Ekman R, Andrade NC, Wharton J, Polak JM, Queiroz e Melo J, Edvinsson L (1993) Peptidergic innervation of human epicardial coronary arteries. Circ Res 73:579–588

Holland PR, Goadsby PJ (2018) Targeted CGRP small molecule antagonists for acute migraine therapy. Neurotherapeutics 15:304–312

Homma S, Kimura T, Sakai S, Yanagi K-I, Miyauchi Y, Aonuma K, Miyauchi T (2014) Calcitonin gene-related peptide protects the myocardium from ischemia induced by endothelin-1: intravital microscopic observation and 31P-MR spectroscopic studies. Life Sci 118:248–254

Humphries KH, Pu A, Gao M, Carere RG, Pilote L (2008) Angina with "normal" coronary arteries: sex differences in outcomes. Am Heart J 155:375–381

Inoue T, Shimizu H, Kaminuma T, Tajima M, Watabe K, Yoshimoto T (1996) Prevention of cerebral vasospasm by calcitonin gene-related peptide slow-release tablet after subarachnoid hemorrhage in monkeys. Neurosurgery 39:984–990

Jianping L, Kevin AC, Donald JD, Scott CS (2013) Renal protective effects of α-calcitonin gene-related peptide in deoxycorticosterone-salt hypertension. Am J Physiol Renal Physiol 304: F1000–F1008

Juul R, Edvinsson L, Gisvold SE, Ekman R, Brubakk AO, Fredriksen TA (1990) Calcitonin gene-related peptide-LI in subarachnoid haemorrhage in man. Signs of activation of the trigemino-cerebrovascular system? Br J Neurosurg 4:171–179

Juul R, Aakhus S, Björnstad K, Gisvold SE, Brubakk AO, Edvinsson L (1994) Calcitonin gene-related peptide (human α-CGRP) counteracts vasoconstriction in human subarachnoid haemorrhage. Neurosci Lett 170:67–70

Kaski JC, Collins P, Nihoyannopoulos P, Maseri A, Poole-Wilson PA, Rosano GMC (1995) Cardiac syndrome X: clinical characteristics and left ventricular function: long-term follow-up study. J Am Coll Cardiol 25:807–814

Keith IM, Tjen-A-Looi S, Kraiczi H, Ekman R (2000) Three-week neonatal hypoxia reduces blood CGRP and causes persistent pulmonary hypertension in rats. Am J Physiol Heart Circ Physiol 279:H1571–H1578

Kurth T, Schürks M, Logroscino G, Buring JE (2009) Migraine frequency and risk of cardiovascular disease in women. Neurology 73:581–588

Labastida-Ramirez A, Rubio-Beltran E, Villalon CM, MaassenVanDenBrink A (2017) Gender aspects of CGRP in migraine. Cephalalgia 333102417739584

Lei J, Zhu F, Zhang Y, Duan L, Lei H, Huang W (2016) Transient receptor potential vanilloid subtype 1 inhibits inflammation and apoptosis via the release of calcitonin gene-related peptide in the heart after myocardial infarction. Cardiology 134:436–443

Lindstedt IH, Edvinsson ML, Evinsson L (2006) Reduced responsiveness of cutaneous microcirculation in essential hypertension – a pilot study. Blood Press 15:275–280

Liu Z, Liu Q, Cai H, Xu C, Liu G, Li Z (2011) Calcitonin gene-related peptide prevents blood–brain barrier injury and brain edema induced by focal cerebral ischemia reperfusion. Regul Pept 171:19–25

Maassen van den Brink A, Rubio-Beltrán E, Duncker D, Villalón CM (2018) Is CGRP receptor blockade cardiovascularly safe? Appropriate studies are needed. Headache 58:1257–1258

MaassenVanDenBrink A, Meijer J, Villalón CM, Ferrari MD (2016) Wiping out CGRP: potential cardiovascular risks. Trends Pharmacol Sci 37:779–788

Mai TH, Wu J, Diedrich A, Garland EM, Robertson D (2014) Calcitonin gene-related peptide (CGRP) in autonomic cardiovascular regulation and vascular structure. J Am Soc Hypertens 8:286–296

McCulloch J, Uddman R, Kingman TA, Edvinsson L (1986) Calcitonin gene-related peptide: functional role in cerebrovascular regulation. Proc Natl Acad Sci 83:5731–5735

Mieres JH, Gulati M, Merz NB, Berman DS, Gerber TC, Hayes SN, Kramer CM, Min JK, Newby LK, Nixon JV, Srichai MB, Pellikka PA, Redberg RF, Wenger NK, Shaw LJ (2014) Role of noninvasive testing in the clinical evaluation of women with suspected ischemic heart disease. Circulation 130:350–379

Mitsikostas DD, Reuter U (2017) Calcitonin gene-related peptide monoclonal antibodies for migraine prevention: comparisons across randomized controlled studies. Curr Opin Neurol 30:272–280

Moskowitz MA, Sakas DE, Wei EP, Kano M, Buzzi MG, Ogilvy C, Kontos HA (1989) Postocclusive cerebral hyperemia is markedly attenuated by chronic trigeminal ganglionectomy. Am J Physiol 257:H1736–H1739

Negro A, Lionetto L, Simmaco M, Martelletti P (2012) CGRP receptor antagonists: an expanding drug class for acute migraine? Expert Opin Investig Drugs 21:807–818

Opgaard OS, Gulbenkian S, Bergdahl A, Barroso CP, Andrade NC, Polak JM, Queiroz e Melo JQ, Edvinsson L (1995) Innervation of human epicardial coronary veins: immunohistochemistry and vasomotility. Cardiovasc Res 29:463–468

Russell FA, King R, Smillie SJ, Kodji X, Brain SD (2014) Calcitonin gene-related peptide: physiology and pathophysiology. Physiol Rev 94:1099

Sacco S, Ornello R, Ripa P, Pistoia F, Carolei A (2013) Migraine and hemorrhagic stroke: a meta-analysis. Stroke 44:3032–3038

Sakas DE, Moskowitz MA, Wei EP, Kontos HA, Kano M, Ogilvy CS (1989) Trigeminovascular fibers increase blood flow in cortical gray matter by axon reflex-like mechanisms during acute severe hypertension or seizures. Proc Natl Acad Sci 86:1401–1405

Schebesch K-M, Herbst A, Bele S, Schödel P, Brawanski A, Stoerr E-M, Lohmeier A, Kagerbauer SM, Martin J, Proescholdt M (2013) Calcitonin-gene related peptide and cerebral vasospasm. J Clin Neurosci 20:584–586

Scher AI, Terwindt GM, Picavet HSJ, Verschuren WMM, Ferrari MD, Launer LJ (2005) Cardio-vascular risk factors and migraine: the GEM population-based study. Neurology 64:614–620

Schurks M, Rist PM, Bigal ME, Buring JE, Lipton RB, Kurth T (2009) Migraine and cardiovascular disease: systematic review and meta-analysis. Br Med J 339:b3914

Schuster NM, Vollbracht S, Rapoport AM (2015) Emerging treatments for the primary headache disorders. Neurol Sci 36(Suppl 1):109–113

Smillie S-J, Brain SD (2011) Calcitonin gene-related peptide (CGRP) and its role in hypertension. Neuropeptides 45:93–104

Smillie S-J, King R, Kodji X, Outzen E, Pozsgai G, Fernandes E, Marshall N, de Winter P, Heads Richard J, Dessapt-Baradez C, Gnudi L, Sams A, Shah Ajay M, Siow Richard C, Brain Susan D (2014) An ongoing role of α-calcitonin gene-related peptide as part of a protective network against hypertension, vascular hypertrophy, and oxidative stress. Hypertension 63:1056–1062

Spector JT, Kahn SR, Jones MR, Jayakumar M, Dalal D, Nazarian S (2010) Migraine headache and ischemic stroke risk: an updated meta-analysis. Am J Med 123:612–624

Stovner LJ, Nichols E, Steiner TJ, Abd-Allah F, Abdelalim A, Al-Raddadi RM, Ansha MG, Barac A, Bensenor IM, Doan LP, Edessa D, Endres M, Foreman KJ, Gankpe FG, Gopalkrishna G, Goulart AC, Gupta R, Hankey GJ, Hay SI, Hegazy MI, Hilawe EH, Kasaeian A, Kassa DH, Khalil I, Khang Y-H, Khubchandan J, Kim YJ, Kokubo Y, Mohammed MA, Moradi-Lakeh M, Nguyen HLT, Nirayo YL, Qorbani M, Ranta A, Roba KT, Safiri S,

Santos IS, Satpathy M, Sawhney M, Shiferaw MS, Shiue I, Smith M, Szoeke CEI, Truong NT, Venketasubramanian N, Weldegwergs KG, Westerman R, Wijeratne T, Tran BX, Yonemoto N, Feigin VL, Vos T, Murray CJL (2018) Global, regional, and national burden of migraine and tension-type headache, 1990–2016: a systematic analysis for the Global Burden of Disease Study 2016. Lancet Neurol 17:954–976

Tepper SJ (2018) History and review of anti-calcitonin gene-related peptide (CGRP) therapies: from translational research to treatment. Headache 58(Suppl 3):238–275

Tzourio C, Tehindrazanarivelo A, Iglesias S, Alperovitch A, Chedru F, d'Anglejan-Chatillon J, Bousser M-G (1995) Case-control study of migraine and risk of ischaemic stroke in young women. Br Med J 310:830–833

Uddman R, Edvinsson L, Ekblad E, Håkanson R, Sundler F (1986) Calcitonin gene-related peptide (CGRP): perivascular distribution and vasodilatory effects. Regul Pept 15:1–23

Valdemarsson S, Edvinsson L, Hedner P, Ekman R (1990) Hormonal influence on calcitonin gene-related peptide in man: effects of sex difference and contraceptive pills. Scand J Clin Lab Invest 50:385–388

Vanmolkot FH, Van Bortel LM, de Hoon JN (2007) Altered arterial function in migraine of recent onset. Neurology 68:1563–1570

Wang LH, Zhou SX, Li RC, Zheng LR, Zhu JH, Hu SJ, Sun YL (2012) Serum levels of calcitonin gene-related peptide and substance P are decreased in patients with diabetes mellitus and coronary artery disease. J Int Med Res 40:134–140

Wang Z, Martorell BC, Wälchli T, Vogel O, Fischer J, Born W, Vogel J (2015) Calcitonin gene-related peptide (CGRP) receptors are important to maintain cerebrovascular reactivity in chronic hypertension. PLoS One 10:e0123697

Wimalawansa SJ, MacIntyre I (1988) Calcitonin gene-related peptide and its specific binding sites in the cardiovascular system of rat. Int J Cardiol 20:29–37

Wrobel Goldberg S, Silberstein SD (2015) Targeting CGRP: a new era for migraine treatment. CNS Drugs 29:443–452

Yadav S, Yadav YS, Goel MM, Singh U, Natu SM, Negi MPS (2014) Calcitonin gene- and parathyroid hormone-related peptides in normotensive and preeclamptic pregnancies: a nested case-control study. Arch Gynecol Obstet 290:897–903

Zhang J-Y, Yan G-T, Liao J, Deng Z-H, Xue H, Wang L-H, Zhang K (2011) Leptin attenuates cerebral ischemia/reperfusion injury partially by CGRP expression. Eur J Pharmacol 671:61–69

CGRP and Painful Pathologies Other than Headache

David A. Walsh and Daniel F. McWilliams

Contents

Abstract

CGRP has long been suspected as a mediator of arthritis pain, although evidence that CGRP directly mediates human musculoskeletal pain remains circumstantial. This chapter describes in depth the evidence surrounding CGRP's association with pain in musculoskeletal disorders and also summarises evidence for CGRP being a direct cause of pain in other conditions. CGRP-immunoreactive nerves are present in musculoskeletal tissues, and CGRP expression is altered in musculoskeletal pain. CGRP modulates musculoskeletal pain through actions both in the periphery and central nervous system. Human observational studies, research

D. A. Walsh (✉)
Pain Centre Versus Arthritis, NIHR Nottingham Biomedical Research Centre and Division of ROD, University of Nottingham, Nottingham, UK

Rheumatology, Sherwood Forest Hospitals NHS Foundation Trust, Nottinghamshire, UK
e-mail: David.walsh@nottingham.ac.uk

D. F. McWilliams
Pain Centre Versus Arthritis, NIHR Nottingham Biomedical Research Centre and Division of ROD, University of Nottingham, Nottingham, UK
e-mail: Dan.mcwilliams@nottingham.ac.uk

© Springer Nature Switzerland AG 2019
S. D. Brain, P. Geppetti (eds.), *Calcitonin Gene-Related Peptide (CGRP) Mechanisms*,
Handbook of Experimental Pharmacology 255, https://doi.org/10.1007/164_2019_242

on animal arthritis models and the few reported randomised controlled trials in humans of treatments that target CGRP provide the context of CGRP as a possible pain biomarker or mediator in conditions other than migraine.

Keywords

Back pain · Central sensitisation · Gastrointestinal · Neck pain · Neuropathic pain · Osteoarthritis · Pain · Pancreatitis · Peripheral sensitisation · Rheumatoid arthritis

Abbreviations

ATF3	Activating transcription factor-3
BDNF	Brain-derived neurotrophic factor
CCL	Chemokine ligand
CGRP	Calcitonin gene-related polypeptide
CNS	Central nervous system
COX2	Cyclooxygenase-2
CRPS	Complex regional pain syndrome
CSF	Cerebrospinal fluid
DRG	Dorsal root ganglion
GFAP	Glial fibrillary acidic protein
GI	Gastrointestinal
IKK-β	Inhibitor of nuclear factor kappa-B kinase subunit beta
MIA	Monoiodoacetate
NGF	Nerve growth factor
NSAID	Non-steroidal anti-inflammatory drug
OA	Osteoarthritis
PGP9.5	Protein gene product 9.5
PKA	Protein kinase-A
RA	Rheumatoid arthritis
RAMP	Receptor activity-modifying protein
TNF	Tumour necrosis factor-alpha
TrkA	Receptor for nerve growth factor
TRPV	Transient receptor potential cation channel subfamily V member
VAS	Visual analogue scale
WOMAC	Western Ontario and McMaster Universities Osteoarthritis Index

1   Introduction

CGRP has long been explored as a neuromodulating peptide in diverse painful conditions, due to its localisation in fine unmyelinated first-order sensory nerves that are activated by nociceptive and painful chemical stimuli. Clinical benefit in migraine may reflect a particular contribution of perivascular CGRP-containing nerves in that condition, either or both due to sensory and efferent vasomotor activities of CGRP. Substantial evidence has been accumulated that CGRP might also be important for musculoskeletal pain and possibly also neuropathic and abdominal pains. To date, however, the translational relevance of these findings remains unproven for treating human pain states other than migraine. More sophisticated models of chronic pain now recognise the importance of peripheral and central sensitisation and contributions from inflammation and the vasculature. Pain is not a single entity, and different treatments might specifically reduce its different components, for example, nociceptive or neuropathic and constant or intermittent. Patient-reported outcomes designed to assess benefits from one class of analgesics might not detect important effects of an analgesic agent acting through different mechanisms. Understanding potential mechanisms by which CGRP might contribute to chronic pain is important in the design of clinical trials to test analgesic efficacy in humans.

This chapter updates previous reviews on CGRP's role in pain (Schou et al. 2017; Walsh et al. 2015). A recent systematic review of the reported associations between CGRP and pain found that outside of migraine and headache translationally robust findings were limited (Schou et al. 2017). Only one randomised controlled trial of CGRP inhibition outside of headache/migraine was identified, and that failed to demonstrate benefit for knee osteoarthritis (OA) pain. However, these findings underestimate CGRP as a target of interest and might underestimate the potential for blocking CGRP to relieve chronic pain.

2   CGRP in Musculoskeletal Pain

Arthritis is the commonest cause of chronic pain, and OA and rheumatoid arthritis (RA) each remains a major burden on both individuals and society. In the UK alone, 7.5 million working days are lost per year due to musculoskeletal conditions. Current treatments often provide incomplete relief of arthritis pain or are associated with important risk of adverse events. Even total joint replacement surgery, one of the most effective treatments to improve quality of life across any condition, leaves a substantial minority of patients with chronic pain and is only appropriate for people with end-stage disease who have often suffered for many years from their arthritis.

CGRP has long been suspected as a mediator of arthritis pain, although evidence that CGRP directly mediates human musculoskeletal pain remains circumstantial. Human joints, bone and muscle are richly innervated with CGRP-immunoreactive sensory nerves, and arthritis has been associated with increased CGRP-like immunoreactivity in joint fluids suggesting increased peripheral CGRP release. Nerve growth factor (NGF) expression is increased in arthritic joints, where it both sensitises peripheral nerves and increases their expression of CGRP. Antibodies to NGF can provide clinically important analgesia in patients with OA. Research in rodents and larger mammals shows that CGRP sensory pathways are remarkably conserved across species, and demonstrate upregulation of CGRP by sensory nerves during arthritis. In animal models, inhibiting CGRP in the CNS or in the joint can reduce arthritis pain behaviour. A more recent clinical trial of antibody blockade of CGRP receptors however did not demonstrate benefit for OA pain in humans (Jin et al. 2018). Does this represent another example of preclinical models not predicting translation to human disease or a failure of clinical trial design to reveal clinically important benefit for human pain? In this chapter, we summarise contributions of CGRP to musculoskeletal pain (Fig. 1) and attempt to interpret apparent inconsistencies in the available evidence.

2.1   Musculoskeletal Innervation by CGRP-Immunoreactive Nerves

CGRP is expressed by fine, unmyelinated sensory nerves supplying articular tissues (Walsh et al. 2015; Dirmeier et al. 2008). In skeletal muscle, CGRP is also present in motoneurons where peripheral release might be myotrophic. Within human synovial joints, CGRP-immunoreactive sensory nerves have been localised to synovium and tendons, ligaments, menisci in the knee and temporomandibular joint as well as periosteum and subchondral bone (Walsh et al. 2015; Dirmeier et al. 2008). Synovial joints studied include knees (Pereira da Silva and Carmo-Fonseca 1990), temporo-mandibular joints and spinal facet joints (Inami et al. 2001; Kallakuri et al. 2004). CGRP-immunoreactive nerves have also been localised to non-synovial components of human sacroiliac joints (Szadek et al. 2008, 2010), muscle (Ohtori et al. 2012), tendon insertions (Spang and Alfredson 2017), intervertebral discs (Ashton et al. 1994; Gruber et al. 2012), vertebral end plates (Brown et al. 1997) and peridural membrane (Bosscher et al. 2016) and in uncovertebral joints of the cervical spine (Brismee et al. 2009). Musculoskeletal innervation by CGRP-immunoreactive nerves has been demonstrated similarly across species in rats (Iwasaki et al. 1995; Ahmed et al. 1993; Hukkanen et al. 1991, 1992a, b; Shinoda et al. 2003), rabbits (Kallakuri et al. 1998) and sheep (Tahmasebi-Sarvestani et al. 1996).

Retrograde neuronal tracing in rodents has demonstrated specific routes through which CGRP-immunoreactive musculoskeletal nerves course before reaching their dorsal root ganglia (DRG). Articular nerves conduct CGRP-immunoreactive fibres from peripheral joints, whereas sympathetic trunks might make important contributions from intervertebral discs (Suseki et al. 1998) or facet joints (Aoki et al. 2004). This is essential information if developing targeted nerve ablation procedures. CGRP-immunoreactive nerves in peripheral joints such as knees or wrists (Kuniyoshi et al. 2007) have their cell bodies in restricted ipsilateral lumbosacral or cervical DRGs. Temporomandibular joints are innervated by CGRP-immunoreactive nerves originating in the ipsilateral trigeminal ganglion. CGRP-immunoreactive nerves from vertebral bodies (Ohtori et al. 2007), facet joints (Ohtori et al. 2000, 2002; Kras et al. 2013; Ishikawa et al. 2005), intervertebral discs or spinal ligaments and sacroiliac joints (Murata et al. 2007) may have cell bodies along a broader range of DRGs, on either side of the body. Dual retrograde labelling experiments suggest that CGRP-immunoreactive axons from knee and lumbar vertebrae might originate from a single DRG cell, perhaps providing one anatomical basis for referred musculoskeletal pain (Ohtori et al. 2003).

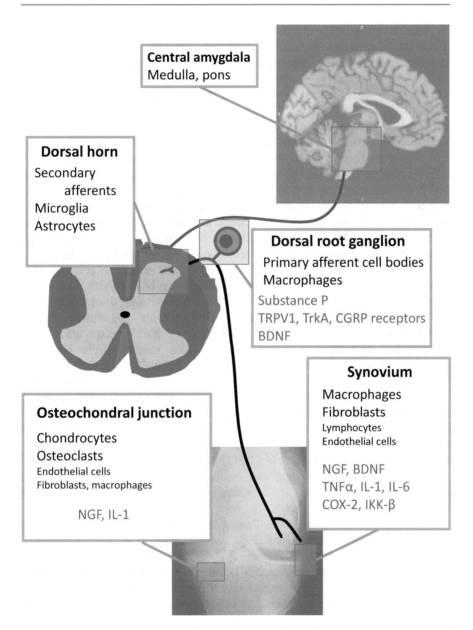

Fig. 1 Arthritis pain mechanisms associated with CGRP. Schematic diagram displaying joint and afferent nervous system structures involved in arthritis pain mechanisms that are associated with CGRP. Nociceptive signals from subchondral bone and synovium are processed within the spinal cord and brain stem. Cytokines and growth factors released within the joint as well as CGRP released from peripheral terminals of articular sensory nerves can sensitise peripheral nociceptive nerves. Altered gene expression in the dorsal root ganglion and increased CGRP release within the dorsal horn modulate activity in secondary afferents. CGRP is widely expressed in the central nervous system, acting in the amygdala, pons and other brainstem regions to augment nociceptive

Arthritis can modify joint innervation, CGRP expression and release (Walsh et al. 2015). In osteoarthritic human joints, CGRP-immunoreactive nerves have been demonstrated in vascular channels within articular cartilage and deep to the outer third of the knee meniscus, structures which are normally avascular and aneural (Ashraf et al. 2011). Osteophytes, regions of new bone formation, also contain CGRP-immunoreactive nerves. Higher densities of CGRP-immunoreactive nerves in the synovium have been associated with human OA (Saxler et al. 2007). Increased CGRP innervation has also been reported at vertebral end plates from people with low back pain (Brown et al. 1997) and in the annulus fibrosus of spondylotic intervertebral discs (Gruber et al. 2012). OA (Ichiseki et al. 2018; Miyamoto et al. 2017), inflammatory arthritis (Shinoda et al. 2003; Weihe et al. 1988; Wu et al. 2002; Kar et al. 1991; Imai et al. 1997a; Ghilardi et al. 2012), myositis (Reinert et al. 1998), inflammation of thoracolumbar fascia (Mense and Hoheisel 2016) and laminectomy (Saxler et al. 2008) in rodents have also been associated with proliferation of CGRP-immunoreactive nerve terminals. This neoinnervation is perhaps in response to local production of neurotrophic factors such as NGF. NGF expression is increased in human synovium (Stoppiello et al. 2014), subchondral bone (Walsh et al. 2010), articular chondrocytes (Iannone et al. 2002) and intervertebral discs (Krock et al. 2014) in painful arthritic conditions. NGF inhibition can reduce CGRP-immunoreactive articular innervation (Ghilardi et al. 2012). Spondylotic human intervertebral discs produce factors which stimulate sensory neurite outgrowth in vitro, an effect which was blocked by antibodies to NGF (Krock et al. 2014).

Higher densities of CGRP-immunoreactive nerves in the synovium have been associated with worse pain in people with OA undergoing arthroplasty (Takano et al. 2017). Similar associations between CGRP innervation density and pain behaviour have been noted in rodent models of OA (Miyamoto et al. 2017) or RA (Ghilardi et al. 2012), thoracolumbar fascial inflammation (Miyagi et al. 2011a) and intervertebral disc injury (Miyagi et al. 2011b, 2013). NGF blockade abrogated the increased CGRP innervation in synovium and reduced pain behaviour in CFA-induced arthritic rats (Ghilardi et al. 2012). Changes in neuroanatomy therefore might contribute to arthritis pain.

Human (Dirmeier et al. 2008; Pereira da Silva and Carmo-Fonseca 1990; Gronblad et al. 1988; Mapp et al. 1990), rat (Murakami et al. 2015; Mapp et al. 1993; Konttinen et al. 1990, 1992; Buma et al. 2000), mouse (Buma et al. 1992) and ovine (Tahmasebi-Sarvestani et al. 2001) arthritis have been associated in some studies with reduced, rather than increased CGRP-immunoreactive nerve densities in synovium. Such apparent denervation might result from synovial hyperplasia

Fig. 1 (continued) signalling. CGRP receptor antagonists with restricted access across the blood-brain barrier have helped elucidate contributions of both peripheral and central CGRP systems to musculoskeletal pain. NGF, nerve growth factor; COX2, cyclooxygenase-2; TNF, tumour necrosis factor-alpha; TRPV, transient receptor potential cation channel subfamily V member; BDNF, brain-derived neurotrophic factor; IKK-β, inhibitor of nuclear factor kappa-B kinase subunit beta; TrkA, receptor for nerve growth factor

outstripping the capacity for nerves to grow into new tissues (Buma et al. 2000). Increased peptide release from nerve terminals might render them invisible to immunohistochemistry for CGRP, despite increased axonal transport of CGRP from the DRG. However, similar reductions in nerve terminal densities detected by immunohistochemistry for constitutive proteins such as protein gene product 9.5 (PGP9.5) suggest that terminal rarefaction is not entirely attributable to peptide release. Nerve terminals might retract from chemically hostile environments, for example, due to the induction of proteases or free radicals during inflammation. CGRP-immunoreactive nerve densities might also be reduced in OA synovium, possibly also associated with synovitis. In animal models, chemical induction of OA by monoiodoacetate (MIA) or surgical induction might directly damage sensory nerves, and increased activating transcription factor-3 (ATF3) in peptidergic DRG cells might indicate nerve injury in rat models of inflammatory arthritis (Nascimento et al. 2011). Early nerve damage might be followed by reinnervation during the repair phase of inflammatory arthritis models (Imai et al. 1997b). Peripheral nerve damage can lead to neuropathic pain, and the sharp and burning pain characteristics described by some people with arthritis (Hochman et al. 2013) might indicate neuropathic pain mechanisms.

2.2 Altered CGRP Expression in Musculoskeletal Pain

Musculoskeletal disease or injury are associated with increased expression of CGRP by sensory ganglia innervating the affected joint (Miyamoto et al. 2017; Kar et al. 1991, 1994; Kuraishi et al. 1989; Nohr et al. 1999; Weihe et al. 1995; Ikeuchi et al. 2009; Hutchins et al. 2000; Hanesch et al. 1993, 1997; Bulling et al. 2001; Carleson et al. 1997; Chen et al. 2008; Ahmed et al. 1995; Damico et al. 2012; Taniguchi et al. 2015; Walker et al. 2000; Donaldson et al. 1992; Smith et al. 1992; Staton et al. 2007; Nieto et al. 2015), intervertebral disc (Lee et al. 2009; Koshi et al. 2010; Kobori et al. 2014) or muscle (Ambalavanar et al. 2006a). Arthritis can induce CGRP expression, although an initial decrease in CGRP immunoreactivity might reflect release of preformed peptide. Arthritis might increase CGRP expression through the production of NGF in the joint. CGRP expression in DRGs is increased by intra-articular injection of NGF (Omae et al. 2015), and NGF inhibition can reduce CGRP expression in DRGs (Iwakura et al. 2010).

Increased CGRP expression by DRGs is maintained for many weeks in rodent models of chronic OA (Ichiseki et al. 2018; Ferland et al. 2011; Ferreira-Gomes et al. 2010; Kawarai et al. 2018), RA (Kuraishi et al. 1989; Staton et al. 2007) or myositis (Ambalavanar et al. 2006b), and this has been associated with increased pain behaviour (Staton et al. 2007; Lee et al. 2009; Ferreira-Gomes et al. 2010; Fernihough et al. 2005). However, increased CGRP expression by DRGs might persist long after remission of Freund's Complete Adjuvant-induced arthritis in rats and after resolution of overt pain behaviour, indicating that increased CGRP expression alone is not a valid biomarker of musculoskeletal pain (Calza et al. 2000). Inhibitory mechanisms, perhaps involving opioids (Calza et al. 2000), might balance

CGRP-induced pain augmentation in this model. Deficiencies in endogenous pain inhibition might explain persistent pain in people with RA whose inflammatory disease is in remission (Walsh and McWilliams 2014). Conversely, genetic depletion of tumour necrosis factor-alpha (TNF) prevented the increase in CGRP in DRGs from mice with MIA-induced OA, but did not affect pain behaviour, suggesting that changes in CGRP expression might not be a major drive to OA pain in this model (Taniguchi et al. 2015).

Inflammation is a major driver of increased CGRP expression by musculoskeletal nerves. Anti-inflammatory treatments such as diclofenac (Kuraishi et al. 1989), the cyclooxygenase-2 (COX2) inhibitor rofecoxib (Staton et al. 2007), corticosteroids (Nohr et al. 1999; Weihe et al. 1995), the TNF blocker etanercept (Horii et al. 2011), inhibition of the inhibitor of nuclear factor kappa-B kinase subunit beta (IKK-β) (Kobori et al. 2014) or the proteasome inhibitor MG132 (Ahmed et al. 2012) each blunt CGRP upregulation in rodent models of arthritis. Sustained nociceptor activation might also contribute to CGRP upregulation in DRGs. Local anaesthetic blunts increased neuropeptide expression in arthritis (Donaldson et al. 1994), as do other strategies that might reduce nociceptive transmission. Increased CGRP expression in spinal tissues of rats with shoulder OA is reduced by mesenchymal stem cell therapy aiming to reduce joint pathology (Ichiseki et al. 2018). Successful intervertebral fusion also abrogates the increased CGRP expression in disc innervation in a rat intervertebral disc injury model (Koshi et al. 2010). Further research is required to determine whether CGRP upregulation is driven by direct effect of inflammatory mediators on CGRP expression or by the increased nociceptive input resulting from peripheral inflammation. Increased expression of CGRP was noted in periodontal inflammation that is not typically painful, suggesting a possible direct stimulus from inflammation (Abd El-Aleem et al. 2004).

Bisphosphonates might also reduce CGRP upregulation and pain in rats with monoiodoacetate (MIA)-induced OA (Yu et al. 2013). Bisphosphonates reduce osteoclast activity but also might have anti-inflammatory actions, so mechanisms by which they reduce pain or CGRP expression remain incompletely understood. The bisphosphonate, alendronic acid, reduced DRG expression of CGRP and pain behaviour in ovariectomised rats without arthritis, suggesting a direct link between bone mineral density and mechanical hyperalgesia (Naito et al. 2017).

2.3   CGRP and Mechanisms of Musculoskeletal Pain

CGRP is released from the central terminals of joint afferents during nociceptive transmission. CGRP immunoreactivity in the dorsal horn (Mapp et al. 1993; Marlier et al. 1991; Sluka and Westlund 1993) and spinal CGRP release (Ichiseki et al. 2018; Nieto et al. 2015; Schaible et al. 1994; Collin et al. 1993; Ogbonna et al. 2013; Puttfarcken et al. 2010; Nanayama et al. 1989) might be increased in animal models of arthritis, although this has not been demonstrated in all models (Malcangio and Bowery 1996). Release exceeding replenishment might sometimes reduce CGRP-immunoreactive nerve densities in the superficial dorsal horn (Kar et al. 1991), in much the same way as described above in peripheral nerve terminals in musculoskeletal tissues. Reduction in spinal CGRP immunoreactivity might be dependent on joint movement (Nakabayashi et al. 2016), consistent with CGRP release during mechanical activation of nociceptors. Reductions in CGRP agonist binding sites in dorsal horn of rats with adjuvant arthritis (Kar et al. 1994) might further represent receptor activation by locally released CGRP. In animal models, spinal release of CGRP might increase soon after the onset of arthritis (Schaible et al. 1994) and persist for many weeks (Collin et al. 1993; Ballet et al. 1998), although release can be tonically suppressed by endogenous opioids, which act on mu and, in arthritis, on delta opioid receptors (Collin et al. 1993; Ballet et al. 1998). CGRP-like immunoreactivity has been detected in cerebrospinal fluid from patients (Lindh et al. 1999) and rats (Carleson et al. 1996) with musculoskeletal pain, although human knee pain was associated with reduced CGRP levels in the cerebrospinal fluid (CSF).

CGRP released into the spinal dorsal horn can modulate second-order neurones and increase nociceptive transmission (Bird et al. 2006). These effects were inhibited by the CGRP receptor antagonist CGRP(8-37), as was spinal gliosis, in collagen-induced arthritis (Nieto et al. 2015). Increased CGRP in the spinal cord induces nocifensive behaviours in response to mechanical stimulation of the joint, an effect which is inhibited by CGRP(8-37), and might be mediated by spinal protein kinase-A (PKA) and by glial fibrillary acidic protein (GFAP) expressing astrocytes (Ogbonna et al. 2013; Cornelison et al. 2016). Intrathecal administration of a CGRP-blocking antibody also reduced hyperalgesia in rats with adjuvant-induced arthritis (Kuraishi et al. 1988). Central sensitisation is a key pain mechanism in complex regional pain syndrome (CRPS). In a mouse model of CRPS following tibial fracture, genetic deficiency of receptor activity-modifying protein-1 (RAMP1), an essential component of the CGRP receptor complex, reduced sensitisation (Li et al. 2018; Shi et al. 2015; Guo et al. 2012). Genetic deletion of CGRP also reduced central sensitisation in a rat knee arthritis model (Zheng et al. 2010).

CGRP might also contribute to pain processing in the brain. Local administration of CGRP receptor antagonists (CGRP(8-37) or BIBN4096BS (olcegepant)) reversed the sensitisation of nociceptive neurons in the central nucleus of the amygdala in anaesthetised arthritic rats (Han et al. 2005). This local CGRP receptor blockade also inhibited hindlimb withdrawal reflexes and ultrasonic vocalisations in awake arthritic rats, the latter possibly indicative of roles of CGRP in affective as well as purely sensory aspects of pain.

Peripheral pain mechanisms involving CGRP might also contribute to musculoskeletal pain. CGRP-like immunoreactivity has been detected in synovial fluids from a range of human joints, including temporomandibular joints (Appelgren et al. 1991, 1995; Holmlund et al. 1991), hips (Wang et al. 2015) and knees (Appelgren et al. 1993; Larsson et al. 1989). Increased peripheral release of CGRP during arthritis might be indicated by local concentrations in synovial fluid which exceed those in plasma (Appelgren et al. 1993). CGRP might also be expressed by non-neuronal tissues including macrophages and fibroblasts within the joint during synovitis (Walsh et al. 2015) and from cells within the intervertebral discs (Ahmed et al. 2019).

CGRP induces sensitisation of joint nociceptors, reducing their thresholds to mechanical stimuli (Bullock et al. 2014). Increased CGRP in the joint can stimulate neuronal and glial expression of proteins implicated in the development of peripheral and central sensitisation including P_2X_3, GFAP and OX-42 (Cady et al. 2011). Osteoarthritic knees appear particularly prone to the sensitising actions of CGRP, and CGRP receptor antagonists administered locally to the joint innervation can reduce this sensitisation and suppress the mechanical hyperalgesia in a rat OA (Bullock et al. 2014). Administration of the peripherally restricted, non-peptide CGRP antagonist, olcegepant (Hirsch et al. 2013), or an antibody directed against CGRP, galcanezumab (LY2951742) (Puttfarcken et al. 2010; Benschop et al. 2014), each reduced weight-bearing asymmetry in rats with OA.

CGRP-immunoreactive nerves in joint tissues are frequently in close association with blood vessels, and CGRP induces synovial vasodilatation. Vascular contributions to musculoskeletal pain have been suggested, but remain unproven (Mapp and Walsh 2012). Reducing blood flow might be expected to increase ischaemic pain in arthritis, although reduced blood flow or angiogenesis by CGRP receptor inhibition might suppress inflammation and thereby reduce inflammatory joint pain (Walsh et al. 2015).

Plasma concentrations of CGRP-like immunoreactivity might represent activity of peripheral sensory nerves. Plasma CGRP levels decreased after treatment of rheumatoid arthritis with etanercept, and this might reflect effective reduction either of pain or of inflammation (Origuchi et al. 2011). Higher concentrations of CGRP in synovial fluids (Appelgren et al. 1995) or blood (Dong et al. 2015a) have been associated with human arthritis pain, but further validation would be required to characterise circulating CGRP as a biomarker of musculoskeletal pain (Schou et al. 2017) or, indeed, of synovitis. Synovial fluid levels of CGRP differed in one study between osteoarthritis and the inflammatory arthritides gout and rheumatoid arthritis, whereas plasma levels showed no differences between groups (Hernanz et al. 1993).

2.4 Disrupting CGRP Pathways to Reduce Musculoskeletal Pain

Several therapeutic strategies might target CGRP and its receptors with the aim of reducing musculoskeletal pain. Sensitisation of CGRP-containing nerves can be reduced by inhibiting arthritis pathology or inhibiting specific sensitising agents. CGRP innervation into musculoskeletal tissues might be interrupted by surgical, physical or chemical means. Pharmacological approaches include small molecule receptor inhibitors, or blocking antibodies targeting CGRP or its receptors. Other strategies might reduce the upregulation of CGRP that is associated with musculoskeletal pathology. Genetic deletion of CGRP or its receptors in animal models provides supportive evidence for the potential of anti-CGRP approaches (Zhang et al. 2001).

As described above, anti-inflammatory treatments including non-steroidal anti-inflammatory drugs (NSAIDs), glucocorticoids or cytokine inhibitors can each reduce neuronal CGRP expression, CGRP peptide levels and nociception in rodent models of arthritis. Inflammation causes joint pain by the sensitisation of articular nerves, in which NGF plays a key role. Inhibiting NGF in the periphery reduces pain in people with knee OA or with low back pain, as demonstrated consistently by randomised controlled trials (Lane et al. 2010; Sanga et al. 2013; Tiseo et al. 2014). NGF displays specificity for peptidergic sensory nerves, strongly implicating contributions to arthritis pain from one or more of their colocalised neurotransmitters or neuromodulators. Clinical trials of neurokinin-1 receptor antagonists in arthritis have not demonstrated useful analgesia (Goldstein et al. 2000). Preclinical evidence that CGRP contributes to peripheral sensitisation in arthritis is consistent with a role for this peptide in the clinical benefit observed with NGF blockade. NGF blocking antibody administration has been associated with rare but important adverse events, including rapidly progressive OA (Hochberg 2015). Alternatives are desirable that might replicate the clinical benefits of NGF blockade while avoiding adverse events.

Local denervation of osteoarthritic joints can also be achieved by cryotherapy or radiofrequency nerve ablation (Oladeji and Cook 2019), and this can be associated with important pain relief. Surgical denervation might also contribute to the analgesic benefit of joint arthroplasty or osteotomy, where subchondral nerves are inevitably sectioned during surgery. Joints receive CGRP fibres through multiple nerve trunks and along blood vessels, such that complete denervation using local procedures might not be possible. Reinnervation can follow denervation procedures, just as regrowth of CGRP-immunoreactive nerves into joint structures can occur during the healing phases of animal arthritis models. Iatrogenic nerve damage has the potential to cause neuropathic pain, which might contribute to persistent pain that sometimes follows joint surgery (Vergne-Salle 2016).

Targeted disruption of CGRP-containing sensory nerves has been achieved using the transient receptor potential cation channel subfamily V member-1 (TRPV1) agonist, capsaicin. Repeated or high-dose capsaicin application can reduce CGRP innervation in joints. Neonatal capsaicin in rodents can induce sustained reductions in articular CGRP immunoreactivity, whereas low doses in adults deplete CGRP from peripheral nerve terminals. Depletion of capsaicin-sensitive nerves in rodents reduced OA-induced pain behaviour (Kalff et al. 2010). Capsaicin injection in osteoarthritic knees has reduced knee pain in a randomised clinical trial (Stevens et al. 2019). Capsaicin injection in rat knees, however, can also induce synovitis, in part due to the release of proinflammatory neuropeptides including CGRP (Mapp et al. 1996). Synovitis has not been reported as an important adverse event with intra-articular capsaicinoids used in clinical trials, possibly due to the careful formulation of this relatively insoluble chemical.

Topical application of capsaicin can locally deplete neuropeptides including CGRP from cutaneous tissues, although penetration into deep joint tissues might be limited. Topical capsaicin was licenced for clinical use following positive randomised controlled trials in knee or hand OA and might also have some benefit for RA pain (Derry et al. 2017). As with other analgesic treatments, approximately half the clinical benefit from topical capsaicin might be attributed to contextual (or placebo) effects (Persson et al. 2018). Full blinding is difficult to achieve in clinical trials due to the temporary burning sensation experienced after capsaicin application.

Pharmacological approaches have now been developed that inhibit activation of CGRP receptors. These include competitive inhibition by small molecules and prevention of receptor engagement using antibodies that bind either CGRP itself or the CGRP receptor. CGRP receptor antagonists can reduce arthritis pain behaviour in rodent models of arthritis (Ogbonna et al. 2013; Cornelison et al. 2016; Kuraishi et al. 1988; Han et al. 2005) or muscle inflammation (Romero-Reyes et al. 2015). Neutralisation of CGRP by the monoclonal antibody galcanezumab dose-dependently reduced pain behaviour as measured by weight-bearing differential in the rat MIA and meniscal tear models of OA pain (Benschop et al. 2014).

Analgesia from CGRP blockade might in part be attributed to reductions in peripheral inflammation resulting from inhibition of CGRP-induced vasodilatation in the joint (Walsh et al. 2015). A CGRP receptor antagonist has also been purported to reduce cartilage degeneration and subchondral bone sclerosis in osteoarthritic mice (Nakasa et al. 2016). Disease modification by CGRP receptor blockade has not, however, been to date demonstrated in human arthritis. Peripheral (intra-articular or closed arterial) administration of CGRP antagonists also reduce nociceptive signalling in joint afferents, suggesting inhibition of CGRP-induced peripheral sensitisation in these animal models (Bullock et al. 2014). Antibodies have very limited penetration across the normal blood-brain (or blood-spinal cord) barrier, and analgesia induced by antibody blockade of CGRP or its receptors in animal models of arthritis might be assumed to be due to effects on the joint rather than directly within the CNS. However, additional contributions of CGRP released in the spinal cord are suggested by reduced hyperalgesic behaviour following intrathecal administration of CGRP receptor antagonist (CGRP(8-37)) in a rat model of inflammatory arthritis (Nieto et al. 2015). Spinal administration of CGRP-blocking antibodies also reduced nociceptive signalling from arthritic joints (Kuraishi et al. 1988). The relative contributions of peripheral and central actions of CGRP to human arthritis pain remain incompletely defined.

A double-blind placebo-controlled trial of 8 weeks of treatment of people with knee OA with the monoclonal antibody to CGRP, galcanezumab, was terminated after interim analysis indicated lack of efficacy in its primary pain outcome (Jin et al. 2018). A celecoxib arm within the same trial showed good improvement (1.2 on a 10 cm visual analogue pain scale), but no dose of galcanezumab was associated with clinically important analgesia above placebo (mean specific analgesic effects were all ≤ 0.5 cm). Clinical development of CGRP-blocking agents has subsequently focussed on migraine rather than arthritis pain (Paemeleire and MaassenVanDenBrink 2018). This single trial however cannot exclude potential benefit from CGRP blockade for people with arthritis. The trial evaluated pain outcomes using visual analogue scale (VAS) and Western Ontario and McMaster Universities Osteoarthritis Index (WOMAC), well-validated patient-reported outcome measures for OA pain. However, given the heterogeneous mechanisms and diverse impacts of arthritis pain, it remains possible that CGRP plays a more important contribution for aspects of pain that are not captured by WOMAC items. In rat OA models, analgesia from CGRP blockade was independent of NSAID-responsive pain and additional to the analgesic benefits of NSAIDs, suggesting that CGRP inhibition and anti-inflammatory treatments might act on different aspects of OA pain (Benschop et al. 2014).

CGRP might particularly contribute to OA pain when there is overactivity of the peripheral peptidergic sensory system. However, post hoc analysis of clinical trial data did not reveal clear prediction of analgesic response by circulating CGRP levels (McNearney et al. 2017), a putative biomarker of high peptidergic nerve activity. However, the study was powered for its primary comparison with placebo rather than for predictive or subgroup analyses. Baseline CGRP levels in OA patients were not associated with WOMAC or VAS pain scores, although they were modestly associated with radiographic OA severity (McNearney et al. 2017).

Lessons might be learned from benefits observed in migraine, where CGRP blockade might be most useful in reducing the frequency and severity of painful episodes, rather than inhibiting ongoing pain. Arthritis pain is similarly episodic, with both intermittent and constant components (Hawker et al. 2008). Patient-reported outcome measures such as ICOAP which capture the important impact of intermittent arthritis pain might be more sensitive to analgesic benefits from CGRP blockade than might be questionnaires that are influenced by constant OA pain, such as WOMAC. Antibodies to CGRP and its receptors have to date had favourable safety profiles, and further exploration of potential benefit for musculoskeletal pain might be justified.

3 CGRP and Pain in Non-musculoskeletal Conditions

CGRP might also be involved in chronic pain states other than migraine and musculoskeletal pain, although human evidence is only observational (Schou et al. 2017). Neuropathic and abdominal pain are major clinical problems, often resistant to existing treatments.

CGRP might contribute to the development and severity of central and peripheral neuropathic pain (Iyengar et al. 2017). Damage to central nociceptive pathways might be traumatic or ischaemic. In an animal model of central neuropathic pain following spinal cord injury, mechanical and thermal allodynia were reduced by intrathecal administration of CGRP(8-37) (Bennett et al. 2000).

Damage to peripheral nerves might be metabolic, as in diabetic neuropathy, traumatic, as in radiculopathy, or iatrogenic as after chemotherapy or surgery. Peripheral neuropathy might be associated both with central and with peripheral sensitisation, mediated by microglial cells, immune cells and immune regulators (Kwiatkowski and Mika 2018). Animals genetically susceptible to neuropathic pain have elevated DRG expression of CGRP (Nitzan-Luques et al. 2013).

Randomised controlled trials of topical capsaicin at sufficient doses to deplete cutaneous nerves of CGRP demonstrate analgesic efficacy above placebo for cutaneous neuropathic pain (Derry et al. 2009). In L5 and L6 lumbar nerve injuries in rats, CGRP was involved in the establishment of hyperalgesia and progression of pain. Intrathecal injections of antagonists (L703,606 and CGRP(8-37)) delayed the induction of mechanical hyperalgesia after nerve ligation (Lee and Kim 2007; Malon et al. 2011). Spinal chemokine ligand (CCL) 5 and p38 might contribute to CGRP-mediated peripheral neuropathy, suggesting inflammatory mechanisms (Malon and Cao 2016). Despite evidence of contributions from CGRP to peripheral neuropathic pain, effects of CGRP receptor antagonists on nerve injury models have shown inconsistencies. Systemic administration of the CGRP receptor antagonist, olcegepant, revealed strong reductions in mechanical allodynia in a model using infraorbital nerve ligation, but no differences were detected using sciatic nerve ligation (Michot et al. 2012). The authors also reported evidence of synergism between CGRP inhibition and naratriptan, a selective 5-HT$_1$ receptor agonist used in migraine, in reducing allodynia following infraorbital nerve ligation, but no synergism was observed after the sciatic nerve ligation (Michot et al. 2012).

CGRP-containing nerves might contribute to gastrointestinal (GI) pain by directly functioning as nociceptive neurones and also by regulating gastrointestinal motility (Evangelista 2014) and inflammation. Inflammatory bowel disorders are often associated with pain, even during remission (Norton et al. 2017; Zielinska et al. 2019). Small and large bowels are densely innervated by CGRP-immunoreactive nerves. Indeed, the major source of CGRP in the gastrointestinal tract is neurons, although neuroendocrine cells also synthesise neuropeptides. CGRP is also found in the diseased gastrointestinal tract (Mozsik et al. 2007) as are CGRP receptors (Cottrell et al. 2012). In people with faecal urgency and incontinence with rectal hypersensitivity, a large increase in CGRP-containing nerve fibre densities was observed, along with other neuropeptides (Chan et al. 2003).

Animal models of GI pain can mimic aspects of inflammation, distension or irritable bowel syndrome and suggest possible roles of CGRP. In models, CGRP nerve density correlates with painful (Qiao and Grider 2009) or hyposensitive (Dong et al. 2015b) phenotypes. Systemic and intrathecal administration of CGRP receptor antagonists each demonstrate analgesic efficacy in GI models (Plourde et al. 1997). Intraperitoneal injection of acetic acid induces immobility and writhing pain behaviours in mice. Targeted genetic deletion of CGRPα from TRPV1-expressing neurons in mice, using CRE-LOX recombination, reduced acetic acid-induced immobility (Spencer et al. 2018). Hypersensitivity to rectal distension in rodents can also be induced by administering acetic acid; and this was completely reversed by CGRP(8-37) (Plourde et al. 1997). Muscle contractions taken as indicative of visceral pain have likewise been inhibited by CGRP(8-37) (Julia and Bueno 1997; Bueno et al. 1997).

People with pancreatitis report that pain is their most common symptom (often referred to the back), pain which is often resistant to treatment (Kuhlmann et al. 2019). Pain mechanisms in pancreatitis might be nociceptive, inflammatory or neuropathic (Kuhlmann et al. 2019), and nociceptive transmission is through the spinal dorsal column (Vera-Portocarrero and Westlund 2005). Experimental pancreatitis can be induced by infusion of trinitobenzene sulfonic acid into the pancreatic duct of rats. In this model, CGRP is upregulated in the DRGs (Winston et al. 2005), and pain sensitivity to electrical stimulation of the pancreas and mechanical stimulation of the abdomen were both reduced by intrathecal administration of CGRP(8-37) (Liu et al. 2011). CGRP upregulation was mediated by NGF (Liu et al. 2011).

In some circumstances, CGRP might display protective actions, and this might be an obstacle to the therapeutic inhibition of CGRP receptors for abdominal pain. CGRP release through activation of TRPV1 appears to be gastroprotective in some models of gastric ulcer (summarised in Evangelista (2014), and the severity of caerulein-induced pancreatitis in rats can be reduced by administration of CGRP during initiation (Warzecha et al. 1997). However, protective effect of CGRP might not be apparent at later stages of pancreatitis when CGRP might exacerbate the condition (Warzecha et al. 2001).

4 Conclusions

Overwhelming evidence implicates CGRP in chronic pain in addition to migraine, even though, to date, this has not translated into clinical benefit from specific CGRP blockade. Musculoskeletal pain has been of major interest, due to its frequency, personal and economic burden and the sparsity of safe and effective treatments. However, chronic neuropathic and gastrointestinal pain might also involve CGRP. The balance between CGRP's involvement in peripheral and central pain mechanisms might be critical to its targeting for non-migraine pain. Peripherally restricted pharmacological agents have advantages of lack of potential for adverse effects within the central nervous system, but their benefit depends on a predominant role of peripheral CGRP receptors rather than those in the dorsal horn of the spinal cord or in the brain. CGRP contributes to sensitisation of peripheral nerves, which we now know makes an important contribution to arthritis pain. Blocking pain caused by sensitisation is attractive in leaving normal protective nociceptive reflexes intact while suppressing pathological pain. Vascular pain might also contribute to musculoskeletal symptoms, although evidence for this remains somewhat circumstantial, and traditional outcome measures might not detect its impact on the patient. Non-selective or selective (e.g. with capsaicin) ablation of musculoskeletal CGRP-containing nerves has entered clinical practice to help relieve arthritis pain. The potential of CGRP-blocking antibodies or receptor antagonists to do likewise deserves further research.

Disclosures DAW has undertaken paid consultancy to Pfizer Ltd., Eli Lilly and Company and GSK Consumer Healthcare. DMcW has been supported by Arthritis Research UK (grant number 20777) and the National Institute for Health Research (NIHR) UK via the NIHR Nottingham Biomedical Research Centre.

References

Abd El-Aleem SA, Morales-Aza BM, Donaldson LF (2004) Sensory neuropeptide mRNA up-regulation is bilateral in periodontitis in the rat: a possible neurogenic component to symmetrical periodontal disease. Eur J Neurosci 19(3):650–658

Ahmed M, Bjurholm A, Kreicbergs A, Schultzberg M (1993) Sensory and autonomic innervation of the facet joint in the rat lumbar spine. Spine 18(14):2121–2126

Ahmed M, Bjurholm A, Schultzberg M, Theodorsson E, Kreicbergs A (1995) Increased levels of substance P and calcitonin gene-related peptide in rat adjuvant arthritis. A combined immuno-histochemical and radioimmunoassay analysis. Arthritis Rheum 38(5):699–709

Ahmed AS, Li J, Erlandsson-Harris H, Stark A, Bakalkin G, Ahmed M (2012) Suppression of pain and joint destruction by inhibition of the proteasome system in experimental osteoarthritis. Pain 153(1):18–26. https://doi.org/10.1016/j.pain.2011.08.001

Ahmed AS, Berg S, Alkass K, Druid H, Hart DA, Svensson CI, Kosek E (2019) NF-kappaB-associated pain-related neuropeptide expression in patients with degenerative disc disease. Int J Mol Sci 20(3):E658. https://doi.org/10.3390/ijms20030658

Ambalavanar R, Dessem D, Moutanni A, Yallampalli C, Yallampalli U, Gangula P, Bai G (2006a) Muscle inflammation induces a rapid increase in calcitonin gene-related peptide (CGRP) mRNA that temporally relates to CGRP immunoreactivity and nociceptive behavior. Neuroscience 143 (3):875–884

Ambalavanar R, Moritani M, Moutanni A, Gangula P, Yallampalli C, Dessem D (2006b) Deep tissue inflammation upregulates neuropeptides and evokes nociceptive behaviors which are modulated by a neuropeptide antagonist. Pain 120(1–2):53–68

Aoki Y, Takahashi Y, Ohtori S, Moriya H, Takahashi K (2004) Distribution and immunocyto-chemical characterization of dorsal root ganglion neurons innervating the lumbar intervertebral disc in rats: a review. Life Sci 74(21):2627–2642

Appelgren A, Appelgren B, Eriksson S, Kopp S, Lundeberg T, Nylander M, Theodorsson E (1991) Neuropeptides in temporomandibular joints with rheumatoid arthritis: a clinical study. Scand J Dent Res 99(6):519–521

Appelgren A, Appelgren B, Kopp S, Lundeberg T, Theodorsson E (1993) Relation between intra-articular temperature of the arthritic temporomandibular joint and presence of calcitonin gene-related peptide in the joint fluid. A clinical study. Acta Odontol Scand 51(5):285–291

Appelgren A, Appelgren B, Kopp S, Lundeberg T, Theodorsson E (1995) Neuropeptides in the arthritic TMJ and symptoms and signs from the stomatognathic system with special consider-ation to rheumatoid arthritis. J Orofac Pain 9(3):215–225

Ashraf S, Wibberley H, Mapp PI, Hill R, Wilson D, Walsh DA (2011) Increased vascular penetration and nerve growth in the meniscus: a potential source of pain in osteoarthritis. Ann Rheum Dis 70(3):523–529. https://doi.org/10.1136/ard.2010.137844

Ashton IK, Roberts S, Jaffray DC, Polak JM, Eisenstein SM (1994) Neuropeptides in the human intervertebral disc. J Orthop Res 12(2):186–192

Ballet S, Mauborgne A, Benoliel JJ, Bourgoin S, Hamon M, Cesselin F, Collin E (1998) Polyarthritis-associated changes in the opioid control of spinal CGRP release in the rat. Brain Res 796(1-2):198–208

Bennett AD, Chastain KM, Hulsebosch CE (2000) Alleviation of mechanical and thermal allodynia by CGRP(8-37) in a rodent model of chronic central pain. Pain 86(1-2):163–175

Benschop RJ, Collins EC, Darling RJ, Allan BW, Leung D, Conner EM, Nelson J, Gaynor B, Xu J, Wang XF, Lynch RA, Li B, McCarty D, Oskins JL, Lin C, Johnson KW, Chambers MG (2014) Development of a novel antibody to calcitonin gene-related peptide for the treatment of osteoarthritis-related pain. Osteoarthr Cartil 22(4):578–585. https://doi.org/10.1016/j.joca.2014.01.009

Bird GC, Han JS, Fu Y, Adwanikar H, Willis WD, Neugebauer V (2006) Pain-related synaptic plasticity in spinal dorsal horn neurons: role of CGRP. Mol Pain 2:31

Bosscher HA, Heavner JE, Grozdanov P, Warraich IA, Wachtel MS, Dertien J (2016) The peridural membrane of the human spine is well innervated. Anat Rec (Hoboken) 299(4):484–491. https://doi.org/10.1002/ar.23315

Brismee JM, Sizer PS Jr, Dedrick GS, Sawyer BG, Smith MP (2009) Immunohistochemical and histological study of human uncovertebral joints: a preliminary investigation. Spine 34 (12):1257–1263. https://doi.org/10.1097/BRS.0b013e31819b2b5d

Brown MF, Hukkanen MV, McCarthy ID, Redfern DR, Batten JJ, Crock HV, Hughes SP, Polak JM (1997) Sensory and sympathetic innervation of the vertebral endplate in patients with degenerative disc disease. J Bone Joint Surg Br 79(1):147–153

Bueno L, Fioramonti J, Delvaux M, Frexinos J (1997) Mediators and pharmacology of visceral sensitivity: from basic to clinical investigations. Gastroenterology 112(5):1714–1743

Bulling DG, Kelly D, Bond S, McQueen DS, Seckl JR (2001) Adjuvant-induced joint inflammation causes very rapid transcription of beta-preprotachykinin and alpha-CGRP genes in innervating sensory ganglia. J Neurochem 77(2):372–382

Bullock CM, Wookey P, Bennett A, Mobasheri A, Dickerson I, Kelly S (2014) Peripheral calcitonin gene-related peptide receptor activation and mechanical sensitization of the joint in rat models of osteoarthritis pain. Arthritis Rheumatol 66(8):2188–2200. https://doi.org/10.1002/art.38656

Buma P, Verschuren C, Versleyen D, Van der Kraan P, Oestreicher AB (1992) Calcitonin gene-related peptide, substance P and GAP-43/B-50 immunoreactivity in the normal and arthrotic knee joint of the mouse. Histochemistry 98(5):327–339

Buma P, Elmans L, Van Den Berg WB, Schrama LH (2000) Neurovascular plasticity in the knee joint of an arthritic mouse model. Anat Rec 260(1):51–61

Cady RJ, Glenn JR, Smith KM, Durham PL (2011) Calcitonin gene-related peptide promotes cellular changes in trigeminal neurons and glia implicated in peripheral and central sensitization. Mol Pain 7:94. https://doi.org/10.1186/1744-8069-7-94

Calza L, Pozza M, Arletti R, Manzini E, Hokfelt T (2000) Long-lasting regulation of galanin, opioid, and other peptides in dorsal root ganglia and spinal cord during experimental polyarthritis. Exp Neurol 164(2):333–343

Carleson J, Alstergren P, Appelgren A, Appelgren B, Kopp S, Srinivasan GR, Theodorsson E, Lundeberg T (1996) Effects of adjuvant on neuropeptide-like immunoreactivity in experimentally induced temporomandibular arthritis in rats. Arch Oral Biol 41(7):705–712

Carleson J, Bileviciute I, Theodorsson E, Appelgren B, Appelgren A, Yousef N, Kopp S, Lundeberg T (1997) Effects of adjuvant on neuropeptide-like immunoreactivity in the temporomandibular joint and trigeminal ganglia. J Orofac Pain 11(3):195–199

Chan CL, Facer P, Davis JB, Smith GD, Egerton J, Bountra C, Williams NS, Anand P (2003) Sensory fibres expressing capsaicin receptor TRPV1 in patients with rectal hypersensitivity and faecal urgency. Lancet 361(9355):385–391

Chen Y, Willcockson HH, Valtschanoff JG (2008) Increased expression of CGRP in sensory afferents of arthritic mice – effect of genetic deletion of the vanilloid receptor TRPV1. Neuropeptides 42(5-6):551–556. https://doi.org/10.1016/j.npep.2008.08.001

Collin E, Mantelet S, Frechilla D, Pohl M, Bourgoin S, Hamon M, Cesselin F (1993) Increased in vivo release of calcitonin gene-related peptide-like material from the spinal cord in arthritic rats. Pain 54(2):203–211

Cornelison LE, Hawkins JL, Durham PL (2016) Elevated levels of calcitonin gene-related peptide in upper spinal cord promotes sensitization of primary trigeminal nociceptive neurons. Neuroscience 339:491–501. https://doi.org/10.1016/j.neuroscience.2016.10.013

Cottrell GS, Alemi F, Kirkland JG, Grady EF, Corvera CU, Bhargava A (2012) Localization of calcitonin receptor-like receptor (CLR) and receptor activity-modifying protein 1 (RAMP1) in human gastrointestinal tract. Peptides 35(2):202–211. https://doi.org/10.1016/j.peptides.2012.03.020

Damico JP, Ervolino E, Torres KR, Sabino Batagello D, Cruz-Rizzolo RJ, Aparecido Casatti C, Arruda Bauer J (2012) Phenotypic alterations of neuropeptide Y and calcitonin gene-related peptide-containing neurons innervating the rat temporomandibular joint during carrageenan-induced arthritis. Eur J Histochem 56(3):e31. https://doi.org/10.4081/ejh.2012.e31

Derry S, Lloyd R, Moore RA, McQuay HJ (2009) Topical capsaicin for chronic neuropathic pain in adults. Cochrane Database Syst Rev 1(4):Cd007393

Derry S, Wiffen PJ, Kalso EA, Bell RF, Aldington D, Phillips T, Gaskell H, Moore RA (2017) Topical analgesics for acute and chronic pain in adults – an overview of Cochrane Reviews. Cochrane Database Syst Rev 5:CD008609. https://doi.org/10.1002/14651858.CD008609.pub2

Dirmeier M, Capellino S, Schubert T, Angele P, Anders S, Straub RH (2008) Lower density of synovial nerve fibres positive for calcitonin gene-related peptide relative to substance P in rheumatoid arthritis but not in osteoarthritis. Rheumatology (Oxford) 47(1):36–40. https://doi.org/10.1093/rheumatology/kem301

Donaldson LF, Harmar AJ, McQueen DS, Seckl JR (1992) Increased expression of preprotachykinin, calcitonin gene-related peptide, but not vasoactive intestinal peptide messenger RNA in dorsal root ganglia during the development of adjuvant monoarthritis in the rat. Brain Res Mol Brain Res 16(1-2):143–149

Donaldson LF, McQueen DS, Seckl JR (1994) Local anaesthesia prevents acute inflammatory changes in neuropeptide messenger RNA expression in rat dorsal root ganglia neurons. Neurosci Lett 175(1-2):111–113

Dong T, Chang H, Zhang F, Chen W, Zhu Y, Wu T, Zhang Y (2015a) Calcitonin gene-related peptide can be selected as a predictive biomarker on progression and prognosis of knee osteoarthritis. Int Orthop 39(6):1237–1243. https://doi.org/10.1007/s00264-015-2744-4

Dong L, Liang X, Sun B, Ding X, Han H, Zhang G, Rong W (2015b) Impairments of the primary afferent nerves in a rat model of diabetic visceral hyposensitivity. Mol Pain 11:74. https://doi.org/10.1186/s12990-015-0075-5

Evangelista S (2014) Capsaicin receptor as target of calcitonin gene-related peptide in the gut. Prog Drug Res 68:259–276

Ferland CE, Laverty S, Beaudry F, Vachon P (2011) Gait analysis and pain response of two rodent models of osteoarthritis. Pharmacol Biochem Behav 97(3):603–610. https://doi.org/10.1016/j.pbb.2010.11.003

Fernihough J, Gentry C, Bevan S, Winter J (2005) Regulation of calcitonin gene-related peptide and TRPV1 in a rat model of osteoarthritis. Neurosci Lett 388(2):75–80. https://doi.org/10.1016/j.neulet.2005.06.044

Ferreira-Gomes J, Adaes S, Sarkander J, Castro-Lopes JM (2010) Phenotypic alterations of neurons that innervate osteoarthritic joints in rats. Arthritis Rheum 62(12):3677–3685. https://doi.org/10.1002/art.27713

Ghilardi JR, Freeman KT, Jimenez-Andrade JM, Coughlin KA, Kaczmarska MJ, Castaneda-Corral G, Bloom AP, Kuskowski MA, Mantyh PW (2012) Neuroplasticity of sensory and sympathetic nerve fibers in a mouse model of a painful arthritic joint. Arthritis Rheum 64(7):2223–2232. https://doi.org/10.1002/art.34385

Goldstein DJ, Wang O, Todd LE, Gitter BD, DeBrota DJ, Iyengar S (2000) Study of the analgesic effect of lanepitant in patients with osteoarthritis pain. Clin Pharmacol Ther 67(4):419–426

Gronblad M, Konttinen YT, Korkala O, Liesi P, Hukkanen M, Polak JM (1988) Neuropeptides in synovium of patients with rheumatoid arthritis and osteoarthritis. J Rheumatol 15(12):1807–1810

Gruber HE, Hoelscher GL, Ingram JA, Hanley EN Jr (2012) Genome-wide analysis of pain-, nerve- and neurotrophin-related gene expression in the degenerating human annulus. Mol Pain 8:63. https://doi.org/10.1186/1744-8069-8-63

Guo TZ, Wei T, Shi X, Li WW, Hou S, Wang L, Tsujikawa K, Rice KC, Cheng K, Clark DJ, Kingery WS (2012) Neuropeptide deficient mice have attenuated nociceptive, vascular, and inflammatory changes in a tibia fracture model of complex regional pain syndrome. Mol Pain 8:85. https://doi.org/10.1186/1744-8069-8-85

Han JS, Li W, Neugebauer V (2005) Critical role of calcitonin gene-related peptide 1 receptors in the amygdala in synaptic plasticity and pain behavior. J Neurosci 25(46):10717–10728

Hanesch U, Pfrommer U, Grubb BD, Schaible HG (1993) Acute and chronic phases of unilateral inflammation in rat's ankle are associated with an increase in the proportion of calcitonin gene-related peptide-immunoreactive dorsal root ganglion cells. Eur J Neurosci 5(2):154–161

Hanesch U, Heppelmann B, Schmidt RF (1997) Quantification of cat's articular afferents containing calcitonin gene-related peptide or substance P innervating normal and acutely inflamed knee joints. Neurosci Lett 233(2–3):105–108

Hawker GA, Davis AM, French MR, Cibere J, Jordan JM, March L, Suarez-Almazor M, Katz JN, Dieppe P (2008) Development and preliminary psychometric testing of a new OA pain measure – an OARSI/OMERACT initiative. Osteoarthr Cartil 16(4):409–414. https://doi.org/10.1016/j.joca.2007.12.015

Hernanz A, De Miguel E, Romera N, Perez-Ayala C, Gijon J, Arnalich F (1993) Calcitonin gene-related peptide II, substance P and vasoactive intestinal peptide in plasma and synovial fluid from patients with inflammatory joint disease. Br J Rheumatol 32(1):31–35

Hirsch S, Corradini L, Just S, Arndt K, Doods H (2013) The CGRP receptor antagonist BIBN4096BS peripherally alleviates inflammatory pain in rats. Pain 154(5):700–707. https://doi.org/10.1016/j.pain.2013.01.002

Hochberg MC (2015) Serious joint-related adverse events in randomized controlled trials of anti-nerve growth factor monoclonal antibodies. Osteoarthr Cartil 23(Suppl 1):S18–S21. https://doi.org/10.1016/j.joca.2014.10.005

Hochman JR, Davis AM, Elkayam J, Gagliese L, Hawker GA (2013) Neuropathic pain symptoms on the modified painDETECT correlate with signs of central sensitization in knee osteoarthritis. Osteoarthritis Cartilage 21(9):1236–1242. https://doi.org/10.1016/j.joca.2013.06.023

Holmlund A, Ekblom A, Hansson P, Lind J, Lundeberg T, Theodorsson E (1991) Concentrations of neuropeptides substance P, neurokinin A, calcitonin gene-related peptide, neuropeptide Y and vasoactive intestinal polypeptide in synovial fluid of the human temporomandibular joint. A correlation with symptoms, signs and arthroscopic findings. Int J Oral Maxillofac Surg 20 (4):228–231

Horii M, Orita S, Nagata M, Takaso M, Yamauchi K, Yamashita M, Inoue G, Eguchi Y, Ochiai N, Kishida S, Aoki Y, Ishikawa T, Arai G, Miyagi M, Kamoda H, Kuniyoshi K, Suzuki M, Nakamura J, Toyone T et al (2011) Direct application of the tumor necrosis factor-alpha inhibitor, etanercept, into a punctured intervertebral disc decreases calcitonin gene-related peptide expression in rat dorsal root ganglion neurons. Spine 36(2):E80–E85. https://doi.org/10.1097/BRS.0b013e3181d4be3c

Hukkanen M, Gronblad M, Rees R, Kottinen YT, Gibson SJ, Hietanen J, Polak JM, Brewerton DA (1991) Regional distribution of mast cells and peptide containing nerves in normal and adjuvant arthritic rat synovium. J Rheumatol 18(2):177–183

Hukkanen M, Konttinen YT, Rees RG, Gibson SJ, Santavirta S, Polak JM (1992a) Innervation of bone from healthy and arthritic rats by substance P and calcitonin gene related peptide containing sensory fibers. J Rheumatol 19(8):1252–1259

Hukkanen M, Konttinen YT, Rees RG, Santavirta S, Terenghi G, Polak JM (1992b) Distribution of nerve endings and sensory neuropeptides in rat synovium, meniscus and bone. Int J Tissue React 14(1):1–10

Hutchins B, Spears R, Hinton RJ, Harper RP (2000) Calcitonin gene-related peptide and substance P immunoreactivity in rat trigeminal ganglia and brainstem following adjuvant-induced inflammation of the temporomandibular joint. Arch Oral Biol 45(4):335–345

Iannone F, De Bari C, Dell'Accio F, Covelli M, Patella V, Lo Bianco G, Lapadula G (2002) Increased expression of nerve growth factor (NGF) and high affinity NGF receptor (p140 TrkA) in human osteoarthritic chondrocytes. Rheumatology (Oxford) 41(12):1413–1418

Ichiseki T, Shimazaki M, Ueda Y, Ueda S, Tsuchiya M, Souma D, Kaneuji A, Kawahara N (2018) Intraarticularly-injected mesenchymal stem cells stimulate anti-inflammatory molecules and inhibit pain related protein and chondrolytic enzymes in a monoiodoacetate-induced rat arthritis model. Int J Mol Sci 19(1):105. https://doi.org/10.3390/ijms19010203

Ikeuchi M, Kolker SJ, Sluka KA (2009) Acid-sensing ion channel 3 expression in mouse knee joint afferents and effects of carrageenan-induced arthritis. J Pain 10(3):336–342. https://doi.org/10.1016/j.jpain.2008.10.010

Imai S, Rauvala H, Konttinen YT, Tokunaga T, Maeda T, Hukuda S, Santavirta S (1997a) Efferent targets of osseous CGRP-immunoreactive nerve fiber before and after bone destruction in adjuvant arthritic rat: an ultramorphological study on their terminal-target relations. J Bone Miner Res 12(7):1018–1027

Imai S, Tokunaga Y, Konttinen YT, Maeda T, Hukuda S, Santavirta S (1997b) Ultrastructure of the synovial sensory peptidergic fibers is distinctively altered in different phases of adjuvant induced arthritis in rats: ultramorphological characterization combined with morphometric and immunohistochemical study for substance P, calcitonin gene related peptide, and protein gene product 9.5. J Rheumatol 24(11):2177–2187

Inami S, Shiga T, Tsujino A, Yabuki T, Okado N, Ochiai N (2001) Immunohistochemical demonstration of nerve fibers in the synovial fold of the human cervical facet joint. J Orthop Res 19(4):593–596

Ishikawa T, Miyagi M, Ohtori S, Aoki Y, Ozawa T, Doya H, Saito T, Moriya H, Takahashi K (2005) Characteristics of sensory DRG neurons innervating the lumbar facet joints in rats. Eur Spine J 14(6):559–564

Iwakura N, Ohtori S, Orita S, Yamashita M, Takahashi K, Kuniyoshi K (2010) Role of low-affinity nerve growth factor receptor inhibitory antibody in reducing pain behavior and calcitonin gene-related Peptide expression in a rat model of wrist joint inflammatory pain. J Hand Surg Am 35 (2):267–273. https://doi.org/10.1016/j.jhsa.2009.10.030

Iwasaki A, Inoue K, Hukuda S (1995) Distribution of neuropeptide-containing nerve fibers in the synovium and adjacent bone of the rat knee joint. Clin Exp Rheumatol 13(2):173–178

Iyengar S, Ossipov MH, Johnson KW (2017) The role of calcitonin gene-related peptide in peripheral and central pain mechanisms including migraine. Pain 158(4):543–559. https://doi.org/10.1097/j.pain.0000000000000831

Jin Y, Smith C, Monteith D, Brown R, Camporeale A, McNearney TA, Deeg MA, Raddad E, Xiao N, de la Pena A, Kivitz AJ, Schnitzer TJ (2018) CGRP blockade by galcanezumab was not associated with reductions in signs and symptoms of knee osteoarthritis in a randomized clinical trial. Osteoarthr Cartil 26(12):1609–1618. https://doi.org/10.1016/j.joca.2018.08.019

Julia V, Bueno L (1997) Tachykininergic mediation of viscerosensitive responses to acute inflammation in rats: role of CGRP. Am J Phys 272(1 Pt 1):G141–G146. https://doi.org/10.1152/ajpgi.1997.272.1.G141

Kalff KM, El Mouedden M, van Egmond J, Veening J, Joosten L, Scheffer GJ, Meert T, Vissers K (2010) Pre-treatment with capsaicin in a rat osteoarthritis model reduces the symptoms of pain and bone damage induced by monosodium iodoacetate. Eur J Pharmacol 641(2–3):108–113. https://doi.org/10.1016/j.ejphar.2010.05.022

Kallakuri S, Cavanaugh JM, Blagoev DC (1998) An immunohistochemical study of innervation of lumbar spinal dura and longitudinal ligaments. Spine 23(4):403–411

Kallakuri S, Singh A, Chen C, Cavanaugh JM (2004) Demonstration of substance P, calcitonin gene-related peptide, and protein gene product 9.5 containing nerve fibers in human cervical facet joint capsules. Spine 29(11):1182–1186

Kar S, Gibson SJ, Rees RG, Jura WG, Brewerton DA, Polak JM (1991) Increased calcitonin gene-related peptide (CGRP), substance P, and enkephalin immunoreactivities in dorsal spinal cord and loss of CGRP-immunoreactive motoneurons in arthritic rats depend on intact peripheral nerve supply. J Mol Neurosci 3(1):7–18

Kar S, Rees RG, Quirion R (1994) Altered calcitonin gene-related peptide, substance P and enkephalin immunoreactivities and receptor binding sites in the dorsal spinal cord of the polyarthritic rat. Eur J Neurosci 6(3):345–354

Kawarai Y, Orita S, Nakamura J, Miyamoto S, Suzuki M, Inage K, Hagiwara S, Suzuki T, Nakajima T, Akazawa T, Ohtori S (2018) Changes in proinflammatory cytokines, neuropeptides, and microglia in an animal model of monosodium iodoacetate-induced hip osteoarthritis. J Orthop Res 36(11):2978–2986. https://doi.org/10.1002/jor.24065

Kobori S, Miyagi M, Orita S, Gemba T, Ishikawa T, Kamoda H, Suzuki M, Hishiya T, Yamada T, Eguchi Y, Arai G, Sakuma Y, Oikawa Y, Aoki Y, Toyone T, Takahashi K, Inoue G, Ohtori S (2014) Inhibiting IkappaB kinase-beta downregulates inflammatory cytokines in injured discs and neuropeptides in dorsal root ganglia innervating injured discs in rats. Spine 39 (15):1171–1177. https://doi.org/10.1097/BRS.0000000000000374

Konttinen YT, Rees R, Hukkanen M, Gronblad M, Tolvanen E, Gibson SJ, Polak JM, Brewerton DA (1990) Nerves in inflammatory synovium: immunohistochemical observations on the adjuvant arthritis rat model. J Rheumatol 17(12):1586–1591

Konttinen YT, Hukkanen M, Segerberg M, Rees R, Kemppinen P, Sorsa T, Saari H, Polak JM, Santavirta S (1992) Relationship between neuropeptide immunoreactive nerves and inflammatory cells in adjuvant arthritic rats. Scand J Rheumatol 21(2):55–59

Koshi T, Ohtori S, Inoue G, Ito T, Yamashita M, Yamauchi K, Suzuki M, Aoki Y, Takahashi K (2010) Lumbar posterolateral fusion inhibits sensory nerve ingrowth into punctured lumbar intervertebral discs and upregulation of CGRP immunoreactive DRG neuron innervating punctured discs in rats. Eur Spine J 19(4):593–600. https://doi.org/10.1007/s00586-009-1237-9

Kras JV, Tanaka K, Gilliland TM, Winkelstein BA (2013) An anatomical and immunohistochemical characterization of afferents innervating the C6-C7 facet joint after painful joint loading in the rat. Spine 38(6):E325–E331. https://doi.org/10.1097/BRS.0b013e318285b5bb

Krock E, Rosenzweig DH, Chabot-Dore AJ, Jarzem P, Weber MH, Ouellet JA, Stone LS, Haglund L (2014) Painful, degenerating intervertebral discs up-regulate neurite sprouting and CGRP through nociceptive factors. J Cell Mol Med 18(6):1213–1225. https://doi.org/10.1111/jcmm.12268

Kuhlmann L, Olesen SS, Olesen AE, Arendt-Nielsen L, Drewes AM (2019) Mechanism-based pain management in chronic pancreatitis – is it time for a paradigm shift? Expert Rev Clin Pharmacol 12(3):249–258. https://doi.org/10.1080/17512433.2019.1571409

Kuniyoshi K, Ohtori S, Ochiai N, Murata R, Matsudo T, Yamada T, Ochiai SS, Moriya H, Takahashi K (2007) Characteristics of sensory DRG neurons innervating the wrist joint in rats. Eur J Pain 11(3):323–328

Kuraishi Y, Nanayama T, Ohno H, Minami M, Satoh M (1988) Antinociception induced in rats by intrathecal administration of antiserum against calcitonin gene-related peptide. Neurosci Lett 92 (3):325–329

Kuraishi Y, Nanayama T, Ohno H, Fujii N, Otaka A, Yajima H, Satoh M (1989) Calcitonin gene-related peptide increases in the dorsal root ganglia of adjuvant arthritic rat. Peptides 10 (2):447–452

Kwiatkowski K, Mika J (2018) The importance of chemokines in neuropathic pain development and opioid analgesic potency. Pharmacol Rep 70(4):821–830. https://doi.org/10.1016/j.pharep.2018.01.006

Lane NE, Schnitzer TJ, Birbara CA, Mokhtarani M, Shelton DL, Smith MD, Brown MT (2010) Tanezumab for the treatment of pain from osteoarthritis of the knee. N Engl J Med 363 (16):1521–1531. https://doi.org/10.1056/NEJMoa0901510

Larsson J, Ekblom A, Henriksson K, Lundeberg T, Theodorsson E (1989) Immunoreactive tachykinins, calcitonin gene-related peptide and neuropeptide Y in human synovial fluid from inflamed knee joints. Neurosci Lett 100(1–3):326–330

Lee SE, Kim JH (2007) Involvement of substance P and calcitonin gene-related peptide in development and maintenance of neuropathic pain from spinal nerve injury model of rat. Neurosci Res 58(3):245–249. https://doi.org/10.1016/j.neures.2007.03.004

Lee M, Kim BJ, Lim EJ, Back SK, Lee JH, Yu SW, Hong SH, Kim JH, Lee SH, Jung WW, Sul D, Na HS (2009) Complete Freund's adjuvant-induced intervertebral discitis as an animal model for discogenic low back pain. Anesth Analg 109(4):1287–1296. https://doi.org/10.1213/ane.0b013e3181b31f39

Li WW, Guo TZ, Shi X, Birklein F, Schlereth T, Kingery WS, Clark JD (2018) Neuropeptide regulation of adaptive immunity in the tibia fracture model of complex regional pain syndrome. J Neuroinflammation 15(1):105. https://doi.org/10.1186/s12974-018-1145-1

Lindh C, Liu Z, Welin M, Ordeberg G, Nyberg F (1999) Low calcitonin gene-related, peptide-like immunoreactivity in cerebrospinal fluid from chronic pain patients. Neuropeptides 33 (6):517–521. https://doi.org/10.1054/npep.1999.0772

Liu L, Shenoy M, Pasricha PJ (2011) Substance P and calcitonin gene related peptide mediate pain in chronic pancreatitis and their expression is driven by nerve growth factor. JOP 12(4):389–394

Malcangio M, Bowery NG (1996) Calcitonin gene-related peptide content, basal outflow and electrically-evoked release from monoarthritic rat spinal cord in vitro. Pain 66(2–3):351–358

Malon JT, Cao L (2016) Calcitonin gene-related peptide contributes to peripheral nerve injury-induced mechanical hypersensitivity through CCL5 and p38 pathways. J Neuroimmunol 297:68–75. https://doi.org/10.1016/j.jneuroim.2016.05.003

Malon JT, Maddula S, Bell H, Cao L (2011) Involvement of calcitonin gene-related peptide and CCL2 production in CD40-mediated behavioral hypersensitivity in a model of neuropathic pain. Neuron Glia Biol 7(2-4):117–128. https://doi.org/10.1017/S1740925X12000026

Mapp PI, Walsh DA (2012) Mechanisms and targets of angiogenesis and nerve growth in osteoarthritis. Nat Rev Rheumatol 8(7):390–398. https://doi.org/10.1038/nrrheum.2012.80

Mapp PI, Kidd BL, Gibson SJ, Terry JM, Revell PA, Ibrahim NB, Blake DR, Polak JM (1990) Substance P-, calcitonin gene-related peptide- and C-flanking peptide of neuropeptide Y-immunoreactive fibres are present in normal synovium but depleted in patients with rheumatoid arthritis. Neuroscience 37(1):143–153

Mapp PI, Terenghi G, Walsh DA, Chen ST, Cruwys SC, Garrett N, Kidd BL, Polak JM, Blake DR (1993) Monoarthritis in the rat knee induces bilateral and time-dependent changes in substance P and calcitonin gene-related peptide immunoreactivity in the spinal cord. Neuroscience 57 (4):1091–1096

Mapp PI, Kerslake S, Brain SD, Blake DR, Cambridge H (1996) The effect of intra-articular capsaicin on nerve fibres within the synovium of the rat knee joint. J Chem Neuroanat 10 (1):11–18

Marlier L, Poulat P, Rajaofetra N, Privat A (1991) Modifications of serotonin-, substance P- and calcitonin gene-related peptide-like immunoreactivities in the dorsal horn of the spinal cord of arthritic rats: a quantitative immunocytochemical study. Exp Brain Res 85(3):482–490

McNearney TA, Smith C, Brown R, Camporeale A, Deeg M, Montieth D, Collins EC, Schnitzer TJ, Kivitz AJ, Talbot J et al (2017) Plasma cgrp concentrations were not associated with patient oa symptoms or response to galcanezumab, a monoclonal antibody against cgrp. Ann Rheum Dis 76:983–984. https://doi.org/10.1136/annrheumdis-2017-eular.2155

Mense S, Hoheisel U (2016) Evidence for the existence of nociceptors in rat thoracolumbar fascia. J Bodyw Mov Ther 20(3):623–628. https://doi.org/10.1016/j.jbmt.2016.01.006

Michot B, Bourgoin S, Viguier F, Hamon M, Kayser V (2012) Differential effects of calcitonin gene-related peptide receptor blockade by olcegepant on mechanical allodynia induced by ligation of the infraorbital nerve vs the sciatic nerve in the rat. Pain 153(9):1939–1948. https://doi.org/10.1016/j.pain.2012.06.009

Miyagi M, Ishikawa T, Kamoda H, Orita S, Kuniyoshi K, Ochiai N, Kishida S, Nakamura J, Eguchi Y, Arai G, Suzuki M, Aoki Y, Toyone T, Takahashi K, Inoue G, Ohtori S (2011a) Assessment of gait in a rat model of myofascial inflammation using the CatWalk system. Spine 36(21):1760–1764. https://doi.org/10.1097/BRS.0b013e3182269732

Miyagi M, Ishikawa T, Orita S, Eguchi Y, Kamoda H, Arai G, Suzuki M, Inoue G, Aoki Y, Toyone T, Takahashi K, Ohtori S (2011b) Disk injury in rats produces persistent increases in pain-related neuropeptides in dorsal root ganglia and spinal cord glia but only transient increases in inflammatory mediators: pathomechanism of chronic diskogenic low back pain. Spine 36 (26):2260–2266. https://doi.org/10.1097/BRS.0b013e31820e68c7

Miyagi M, Ishikawa T, Kamoda H, Suzuki M, Sakuma Y, Orita S, Oikawa Y, Aoki Y, Toyone T, Takahashi K, Inoue G, Ohtori S (2013) Assessment of pain behavior in a rat model of intervertebral disc injury using the CatWalk gait analysis system. Spine 38(17):1459–1465. https://doi.org/10.1097/BRS.0b013e318299536a

Miyamoto S, Nakamura J, Ohtori S, Orita S, Nakajima T, Omae T, Hagiwara S, Takazawa M, Suzuki M, Suzuki T, Takahashi K (2017) Pain-related behavior and the characteristics of dorsal-root ganglia in a rat model of hip osteoarthritis induced by mono-iodoacetate. J Orthop Res 35 (7):1424–1430. https://doi.org/10.1002/jor.23395

Mozsik G, Szolcsanyi J, Domotor A (2007) Capsaicin research as a new tool to approach of the human gastrointestinal physiology, pathology and pharmacology. Inflammopharmacology 15 (6):232–245. https://doi.org/10.1007/s10787-007-1584-2

Murakami K, Nakagawa H, Nishimura K, Matsuo S (2015) Changes in peptidergic fiber density in the synovium of mice with collagenase-induced acute arthritis. Can J Physiol Pharmacol 93 (6):435–441. https://doi.org/10.1139/cjpp-2014-0446

Murata Y, Takahashi K, Ohtori S, Moriya H (2007) Innervation of the sacroiliac joint in rats by calcitonin gene-related peptide-immunoreactive nerve fibers and dorsal root ganglion neurons. Clin Anat 20(1):82–88

Naito Y, Wakabayashi H, Kato S, Nakagawa T, Iino T, Sudo A (2017) Alendronate inhibits hyperalgesia and suppresses neuropeptide markers of pain in a mouse model of osteoporosis. J Orthop Sci 22(4):771–777. https://doi.org/10.1016/j.jos.2017.02.001

Nakabayashi K, Sakamoto J, Kataoka H, Kondo Y, Hamaue Y, Honda Y, Nakano J, Okita M (2016) Effect of continuous passive motion initiated after the onset of arthritis on inflammation and secondary hyperalgesia in rats. Physiol Res 65(4):683–691

Nakasa T, Ishikawa M, Takada T, Miyaki S, Ochi M (2016) Attenuation of cartilage degeneration by calcitonin gene-related peptide receptor antagonist via inhibition of subchondral bone sclerosis in osteoarthritis mice. J Orthop Res 34(7):1177–1184. https://doi.org/10.1002/jor.23132

Nanayama T, Kuraishi Y, Ohno H, Satoh M (1989) Capsaicin-induced release of calcitonin gene-related peptide from dorsal horn slices is enhanced in adjuvant arthritic rats. Neurosci Res 6 (6):569–572

Nascimento D, Pozza DH, Castro-Lopes JM, Neto FL (2011) Neuronal injury marker ATF-3 is induced in primary afferent neurons of monoarthritic rats. Neurosignals 19(4):210–221. https://doi.org/10.1159/000330195

Nieto FR, Clark AK, Grist J, Chapman V, Malcangio M (2015) Calcitonin gene-related peptide-expressing sensory neurons and spinal microglial reactivity contribute to pain states in collagen-induced arthritis. Arthritis Rheumatol 67(6):1668–1677. https://doi.org/10.1002/art.39082

Nitzan-Luques A, Minert A, Devor M, Tal M (2013) Dynamic genotype-selective "phenotypic switching" of CGRP expression contributes to differential neuropathic pain phenotype. Exp Neurol 250:194–204. https://doi.org/10.1016/j.expneurol.2013.09.011

Nohr D, Schafer MK, Persson S, Romeo H, Nyberg F, Post C, Ekstrom G, Weihe E (1999) Calcitonin gene-related peptide gene expression in collagen-induced arthritis is differentially regulated in primary afferents and motoneurons: influence of glucocorticoids. Neuroscience 93 (2):759–773

Norton C, Czuber-Dochan W, Artom M, Sweeney L, Hart A (2017) Systematic review: interventions for abdominal pain management in inflammatory bowel disease. Aliment Pharmacol Ther 46(2):115–125. https://doi.org/10.1111/apt.14108

Ogbonna AC, Clark AK, Gentry C, Hobbs C, Malcangio M (2013) Pain-like behaviour and spinal changes in the monosodium iodoacetate model of osteoarthritis in C57Bl/6 mice. Eur J Pain 17 (4):514–526. https://doi.org/10.1002/j.1532-2149.2012.00223.x

Ohtori S, Takahashi K, Chiba T, Yamagata M, Sameda H, Moriya H (2000) Substance P and calcitonin gene-related peptide immunoreactive sensory DRG neurons innervating the lumbar facet joints in rats. Auton Neurosci 86(1-2):13–17

Ohtori S, Moriya H, Takahashi K (2002) Calcitonin gene-related peptide immunoreactive sensory DRG neurons innervating the cervical facet joints in rats. J Orthop Sci 7(2):258–261

Ohtori S, Takahashi K, Chiba T, Yamagata M, Sameda H, Moriya H (2003) Calcitonin gene-related peptide immunoreactive neurons with dichotomizing axons projecting to the lumbar muscle and knee in rats. Eur Spine J 12(6):576–580

Ohtori S, Inoue G, Koshi T, Ito T, Yamashita M, Yamauchi K, Suzuki M, Doya H, Moriya H, Takahashi Y, Takahashi K (2007) Characteristics of sensory dorsal root ganglia neurons innervating the lumbar vertebral body in rats. J Pain 8(6):483–488

Ohtori S, Miyagi M, Takaso M, Inoue G, Orita S, Eguchi Y, Ochiai N, Kishida S, Kuniyoshi K, Nakamura J, Aoki Y, Ishikawa T, Arai G, Kamoda H, Suzuki M, Toyone T, Takahashi K (2012) Differences in damage to CGRP immunoreactive sensory nerves after two lumbar surgical approaches: investigation using humans and rats. Spine 37(3):168–173. https://doi.org/10.1097/BRS.0b013e31821258f7

Oladeji LO, Cook JL (2019) Cooled radio frequency ablation for the treatment of osteoarthritis-related knee pain: evidence, indications, and outcomes. J Knee Surg 32(1):65–71. https://doi.org/10.1055/s-0038-1675418

Omae T, Nakamura J, Ohtori S, Orita S, Yamauchi K, Miyamoto S, Hagiwara S, Kishida S, Takahashi K (2015) A novel rat model of hip pain by intra-articular injection of nerve growth factor-characteristics of sensory innervation and inflammatory arthritis. Mod Rheumatol 25(6):931–936. https://doi.org/10.3109/14397595.2015.1023977

Origuchi T, Iwamoto N, Kawashiri SY, Fujikawa K, Aramaki T, Tamai M, Arima K, Nakamura H, Yamasaki S, Ida H, Kawakami A, Ueki Y, Matsuoka N, Nakashima M, Mizokami A, Kawabe Y, Mine M, Fukuda T, Eguchi K (2011) Reduction in serum levels of substance P in patients with rheumatoid arthritis by etanercept, a tumor necrosis factor inhibitor. Mod Rheumatol 21(3):244–250. https://doi.org/10.1007/s10165-010-0384-5

Paemeleire K, MaassenVanDenBrink A (2018) Calcitonin-gene-related peptide pathway mAbs and migraine prevention. Curr Opin Neurol 31(3):274–280. https://doi.org/10.1097/WCO.0000000000000548

Pereira da Silva JA, Carmo-Fonseca M (1990) Peptide containing nerves in human synovium: immunohistochemical evidence for decreased innervation in rheumatoid arthritis. J Rheumatol 17(12):1592–1599

Persson MSM, Stocks J, Walsh DA, Doherty M, Zhang W (2018) The relative efficacy of topical non-steroidal anti-inflammatory drugs and capsaicin in osteoarthritis: a network meta-analysis of randomised controlled trials. Osteoarthr Cartil 26(12):1575–1582. https://doi.org/10.1016/j.joca.2018.08.008

Plourde V, St-Pierre S, Quirion R (1997) Calcitonin gene-related peptide in viscerosensitive response to colorectal distension in rats. Am J Phys 273(1 Pt 1):G191–G196. https://doi.org/10.1152/ajpgi.1997.273.1.G191

Puttfarcken PS, Han P, Joshi SK, Neelands TR, Gauvin DM, Baker SJ, Lewis LG, Bianchi BR, Mikusa JP, Koenig JR, Perner RJ, Kort ME, Honore P, Faltynek CR, Kym PR, Reilly RM (2010) A-995662 [(R)-8-(4-methyl-5-(4-(trifluoromethyl)phenyl)oxazol-2-ylamino)-1,2,3,4-tetrahydr onaphthalen-2-ol], a novel, selective TRPV1 receptor antagonist, reduces spinal release of glutamate and CGRP in a rat knee joint pain model. Pain 150(2):319–326. https://doi.org/10.1016/j.pain.2010.05.015

Qiao LY, Grider JR (2009) Colitis induces calcitonin gene-related peptide expression and Akt activation in rat primary afferent pathways. Exp Neurol 219(1):93–103. https://doi.org/10.1016/j.expneurol.2009.04.026

Reinert A, Kaske A, Mense S (1998) Inflammation-induced increase in the density of neuropeptide-immunoreactive nerve endings in rat skeletal muscle. Exp Brain Res 121(2):174–180

Romero-Reyes M, Pardi V, Akerman S (2015) A potent and selective calcitonin gene-related peptide (CGRP) receptor antagonist, MK-8825, inhibits responses to nociceptive trigeminal activation: role of CGRP in orofacial pain. Exp Neurol 271:95–103. https://doi.org/10.1016/j.expneurol.2015.05.005

Sanga P, Katz N, Polverejan E, Wang S, Kelly KM, Haeussler J, Thipphawong J (2013) Efficacy, safety, and tolerability of fulranumab, an anti-nerve growth factor antibody, in the treatment of patients with moderate to severe osteoarthritis pain. Pain 154(10):1910–1919. https://doi.org/10.1016/j.pain.2013.05.051

Saxler G, Loer F, Skumavc M, Pfortner J, Hanesch U (2007) Localization of SP- and CGRP-immunopositive nerve fibers in the hip joint of patients with painful osteoarthritis and of patients with painless failed total hip arthroplasties. Eur J Pain 11(1):67–74. https://doi.org/10.1016/j.ejpain.2005.12.011

Saxler G, Brankamp J, von Knoch M, Loer F, Hilken G, Hanesch U (2008) The density of nociceptive SP- and CGRP-immunopositive nerve fibers in the dura mater lumbalis of rats is enhanced after laminectomy, even after application of autologous fat grafts. Eur Spine J 17 (10):1362–1372. https://doi.org/10.1007/s00586-008-0741-7

Schaible HG, Freudenberger U, Neugebauer V, Stiller RU (1994) Intraspinal release of immunoreactive calcitonin gene-related peptide during development of inflammation in the joint in vivo – a study with antibody microprobes in cat and rat. Neuroscience 62(4):1293–1305

Schou WS, Ashina S, Amin FM, Goadsby PJ, Ashina M (2017) Calcitonin gene-related peptide and pain: a systematic review. J Headache Pain 18(1):34. https://doi.org/10.1186/s10194-017-0741-2

Shi X, Guo TZ, Wei T, Li WW, Clark DJ, Kingery WS (2015) Facilitated spinal neuropeptide signaling and upregulated inflammatory mediator expression contribute to postfracture nociceptive sensitization. Pain 156(10):1852–1863. https://doi.org/10.1097/j.pain.0000000000000204

Shinoda M, Honda T, Ozaki N, Hattori H, Mizutani H, Ueda M, Sugiura Y (2003) Nerve terminals extend into the temporomandibular joint of adjuvant arthritic rats. Eur J Pain 7(6):493–505

Sluka KA, Westlund KN (1993) Behavioral and immunohistochemical changes in an experimental arthritis model in rats. Pain 55(3):367–377

Smith GD, Harmar AJ, McQueen DS, Seckl JR (1992) Increase in substance P and CGRP, but not somatostatin content of innervating dorsal root ganglia in adjuvant monoarthritis in the rat. Neurosci Lett 137(2):257–260

Spang C, Alfredson H (2017) Richly innervated soft tissues covering the superficial aspect of the extensor origin in patients with chronic painful tennis elbow – implication for treatment? J Musculoskelet Neuronal Interact 17(2):97–103

Spencer NJ, Magnusdottir EI, Jakobsson JET, Kestell G, Chen BN, Morris D, Brookes SJ, Lagerstrom MC (2018) CGRPalpha within the Trpv1-Cre population contributes to visceral nociception. Am J Physiol Gastrointest Liver Physiol 314(2):G188–G200. https://doi.org/10.1152/ajpgi.00188.2017

Staton PC, Wilson AW, Bountra C, Chessell IP, Day NC (2007) Changes in dorsal root ganglion CGRP expression in a chronic inflammatory model of the rat knee joint: differential modulation by rofecoxib and paracetamol. Eur J Pain 11(3):283–289

Stevens RM, Ervin J, Nezzer J, Nieves Y, Guedes K, Burges R, Hanson PD, Campbell JN (2019) Randomized, double-blind, placebo-controlled trial of intra-articular CNTX-4975 (trans-capsaicin) for pain associated with osteoarthritis of the knee. Arthritis Rheumatol. https://doi.org/10.1002/art.40894

Stoppiello LA, Mapp PI, Wilson D, Hill R, Scammell BE, Walsh DA (2014) Structural associations of symptomatic knee osteoarthritis. Arthritis Rheumatol 66(11):3018–3027. https://doi.org/10.1002/art.38778

Suseki K, Takahashi Y, Takahashi K, Chiba T, Yamagata M, Moriya H (1998) Sensory nerve fibres from lumbar intervertebral discs pass through rami communicantes. A possible pathway for discogenic low back pain. J Bone Joint Surg Br 80(4):737–742

Szadek KM, Hoogland PV, Zuurmond WW, de Lange JJ, Perez RS (2008) Nociceptive nerve fibers in the sacroiliac joint in humans. Reg Anesth Pain Med 33(1):36–43

Szadek KM, Hoogland PV, Zuurmond WW, De Lange JJ, Perez RS (2010) Possible nociceptive structures in the sacroiliac joint cartilage: an immunohistochemical study. Clin Anat 23 (2):192–198. https://doi.org/10.1002/ca.20908

Tahmasebi-Sarvestani A, Tedman RA, Goss A (1996) Neural structures within the sheep temporomandibular joint. J Orofac Pain 10(3):217–231

Tahmasebi-Sarvestani A, Tedman R, Goss AN (2001) The influence of experimentally induced osteoarthrosis on articular nerve fibers of the sheep temporomandibular joint. J Orofac Pain 15 (3):206–217

Takano S, Uchida K, Inoue G, Minatani A, Miyagi M, Aikawa J, Iwase D, Onuma K, Mukai M, Takaso M (2017) Increase and regulation of synovial calcitonin gene-related peptide expression in patients with painful knee osteoarthritis. J Pain Res 10:1099–1104. https://doi.org/10.2147/JPR.S135939

Taniguchi A, Ishikawa T, Miyagi M, Kamoda H, Sakuma Y, Oikawa Y, Kubota G, Inage K, Sainoh T, Nakamura J, Aoki Y, Toyone T, Inoue G, Suzuki M, Yamauchi K, Suzuki T, Takahashi K, Ohtori S, Orita S (2015) Decreased calcitonin gene-related peptide expression in the dorsal root ganglia of TNF-deficient mice in a monoiodoacetate-induced knee osteoarthritis model. Int J Clin Exp Pathol 8(10):12967–12971

Tiseo PJ, Kivitz AJ, Ervin JE, Ren H, Mellis SJ (2014) Fasinumab (REGN475), an antibody against nerve growth factor for the treatment of pain: results from a double-blind, placebo-controlled exploratory study in osteoarthritis of the knee. Pain 155(7):1245–1252. https://doi.org/10.1016/j.pain.2014.03.018

Vera-Portocarrero L, Westlund KN (2005) Role of neurogenic inflammation in pancreatitis and pancreatic pain. Neurosignals 14(4):158–165. https://doi.org/10.1159/000087654

Vergne-Salle P (2016) Management of neuropathic pain after knee surgery. Joint Bone Spine 83 (6):657–663. https://doi.org/10.1016/j.jbspin.2016.06.001

Walker JS, Scott C, Bush KA, Kirkham BW (2000) Effects of the peripherally selective kappa-opioid asimadoline, on substance P and CGRP mRNA expression in chronic arthritis of the rat. Neuropeptides 34(3–4):193–202. https://doi.org/10.1054/npep.2000.0813

Walsh DA, McWilliams DF (2014) Mechanisms, impact and management of pain in rheumatoid arthritis. Nat Rev Rheumatol 10(10):581–592. https://doi.org/10.1038/nrrheum.2014.64

Walsh DA, McWilliams DF, Turley MJ, Dixon MR, Franses RE, Mapp PI, Wilson D (2010) Angiogenesis and nerve growth factor at the osteochondral junction in rheumatoid arthritis and osteoarthritis. Rheumatology (Oxford) 49(10):1852–1861. https://doi.org/10.1093/rheumatology/keq188

Walsh DA, Mapp PI, Kelly S (2015) Calcitonin gene-related peptide in the joint: contributions to pain and inflammation. Br J Clin Pharmacol 80(5):965–978. https://doi.org/10.1111/bcp.12669

Wang H, Zhang X, He JY, Zheng XF, Li D, Li Z, Zhu JF, Shen C, Cai GQ, Chen XD (2015) Increasing expression of substance P and calcitonin gene-related peptide in synovial tissue and fluid contribute to the progress of arthritis in developmental dysplasia of the hip. Arthritis Res Ther 17:4. https://doi.org/10.1186/s13075-014-0513-1

Warzecha Z, Dembinski A, Ceranowicz P, Konturek PC, Stachura J, Konturek SJ, Niemiec J (1997) Protective effect of calcitonin gene-related peptide against caerulein-induced pancreatitis in rats. J Physiol Pharmacol 48(4):775–787

Warzecha Z, Dembinski A, Ceranowicz P, Stachura J, Tomaszewska R, Konturek SJ (2001) Effect of sensory nerves and CGRP on the development of caerulein-induced pancreatitis and pancreatic recovery. J Physiol Pharmacol 52(4 Pt 1):679–704

Weihe E, Nohr D, Millan MJ, Stein C, Muller S, Gramsch C, Herz A (1988) Peptide neuroanatomy of adjuvant-induced arthritic inflammation in rat. Agents Actions 25(3–4):255–259

Weihe E, Nohr D, Schafer MK, Persson S, Ekstrom G, Kallstrom J, Nyberg F, Post C (1995) Calcitonin gene related peptide gene expression in collagen-induced arthritis. Can J Physiol Pharmacol 73(7):1015–1019

Winston JH, He ZJ, Shenoy M, Xiao SY, Pasricha PJ (2005) Molecular and behavioral changes in nociception in a novel rat model of chronic pancreatitis for the study of pain. Pain 117 (1-2):214–222. https://doi.org/10.1016/j.pain.2005.06.013

Wu Z, Nagata K, Iijima T (2002) Involvement of sensory nerves and immune cells in osteophyte formation in the ankle joint of adjuvant arthritic rats. Histochem Cell Biol 118(3):213–220

Yu D, Liu F, Liu M, Zhao X, Wang X, Li Y, Mao Y, Zhu Z (2013) The inhibition of subchondral bone lesions significantly reversed the weight-bearing deficit and the overexpression of CGRP in DRG neurons, GFAP and Iba-1 in the spinal dorsal horn in the monosodium iodoacetate induced model of osteoarthritis pain. PLoS One 8(10):e77824. https://doi.org/10.1371/journal.pone.0077824

Zhang L, Hoff AO, Wimalawansa SJ, Cote GJ, Gagel RF, Westlund KN (2001) Arthritic calcitonin/ alpha calcitonin gene-related peptide knockout mice have reduced nociceptive hypersensitivity. Pain 89(2–3):265–273

Zheng S, Li W, Xu M, Bai X, Zhou Z, Han J, Shyy JY, Wang X (2010) Calcitonin gene-related peptide promotes angiogenesis via AMP-activated protein kinase. Am J Physiol Cell Physiol 299(6):C1485–C1492. https://doi.org/10.1152/ajpcell.00173.2010

Zielinska A, Salaga M, Wlodarczyk M, Fichna J (2019) Focus on current and future management possibilities in inflammatory bowel disease-related chronic pain. Int J Color Dis 34(2):217–227. https://doi.org/10.1007/s00384-018-3218-0

Calcitonin Gene-Related Peptide Antagonists and Therapeutic Antibodies

Roxana-Maria Rujan and Christopher A. Reynolds

Contents

Abstract

The calcitonin gene-related peptide (CGRP) receptor is composed of the calcitonin receptor-like receptor (CLR, a class B GPCR) and a single-pass membrane protein known as receptor activity modifying protein type 1 (RAMP1). The levels of the CGRP peptide increase during a migraine attack and infusion of CGRP can provoke a migraine attack. Consequently, there is much interest in inhibiting the

R.-M. Rujan · C. A. Reynolds (✉)
School of Biological Sciences, University of Essex, Colchester, UK
e-mail: reync@essex.ac.uk

© Springer Nature Switzerland AG 2018
S. D. Brain, P. Geppetti (eds.), *Calcitonin Gene-Related Peptide (CGRP) Mechanisms*, Handbook of Experimental Pharmacology 255, https://doi.org/10.1007/164_2018_173

actions of CGRP as a way to control migraine. Here we describe the development of small molecule antagonists designed to bind to the CGRP receptor to block its action by preventing binding of the CGRP peptide. We also describe the development of antibody drugs, designed to bind either to the CGRP receptor to block its action, or to bind directly to the CGRP peptide. The field has been very active, with one antibody drug approved and three antibody drugs in phase III clinical trial. Initial programs on the development CGRP antagonists were frustrated by liver toxicity but the current outlook is very promising with five small molecule antagonists in various stages of clinical trial.

Keywords

Antibody drugs · Calcitonin-receptor-like receptor · CGRP antagonist · CGRP peptide · Migraine

1   Background

The calcitonin gene-related peptide (CGRP) receptor is composed of the calcitonin receptor-like receptor (CLR), a member of the class B GPCR family, and a single-pass membrane protein known as receptor activity-modifying protein type 1 (RAMP1), involved in modulation of hormone selectivity (Poyner et al. 2002; Pal et al. 2012). Detailed studies have also shown that a new intracellular peripheral membrane protein known as CGRP-receptor component protein (RCP) is required to enable signal transduction (Dickerson 2013).

The cognate ligand of the CGRP receptor, CGRP, consists of 37 amino acids that present a disulfide-bonded ring at positions 2 and 7 of its N-terminus which plays an important part in receptor activation (Barwell et al. 2013; Liang et al. 2018). In addition, C-terminal amidation of the peptide plays a key part in ligand-receptor interaction (O'Connell et al. 1993; Hay and Walker 2017). It has been shown that there are actually two types of CGRP, αCGRP and βCGRP, which in humans differ only by three amino acids but share similar activities (Hay and Walker 2017). The predominant form, αCGRP, is a result of the alternative splicing of the calcitonin (CT) gene *CALCA*, while βCGRP is a transcription product of its own gene, *CALCB*, which shares a high homology to the CT gene (Poyner et al. 2002; Hay and Walker 2017). However, in this review the term CGRP will be used without differentiating the two forms of the peptide, except if it is important. CGRP is widely expressed in the central and peripheral nervous system, including the trigeminovascular pathways, and is consistent with modulation of vasodilatation and transmission of nociceptive information (Deen et al. 2017). CGRP and its receptor are also present in the cardiovascular system where they play a protective role (Deen et al. 2017). During migraine, CGRP is released from the trigeminovascular system. At peripheral synapses, CGRP release is associated with vasodilatation on the smooth muscle cells of meningeal and cerebral blood vessels (Deen et al. 2017). At central synapses, it has been assumed that CGRP release is associated with pain transmission via the brain stem and midbrain to the thalamus and higher cortical pain regions (Eftekhari and Edvinsson 2011). Several studies also revealed that the level of CGRP increases during a migraine attack and infusion of CGRP can provoke this kind of attack in predisposed individuals (Durham and Vause 2010).

To this end, different approaches are being investigated to target CGRP or its receptor to diminish their activity and hence to prevent or treat a migraine attack. Interestingly, a functional antagonist of the CGRP receptor ($CGRP_{8-37}$) can be obtained by deletion of the first seven residues of the CGRP, which are important for receptor activation (Bell 2014). It has been shown that $CGRP_{8-37}$ inhibits vasodilatation and neurogenic inflammation in animal models, but was not useful in clinical studies because its short half-life adds to the lack of potency in vivo (Durham and Vause 2010). However, information gained from these studies supported the development of non-peptide molecules that can block the activity of the peptide at its receptor (Durham and Vause 2010). Therefore, this review aims to focus on the mechanism of action of existing CGRP antagonists and antibodies.

2   CGRP Antagonists

2.1   Olcegepant (BIBN4096BS)

Olcegepant or BIBN4096BS was the first non-peptide antagonist that presented a very high affinity and specificity for the CGRP receptor and that inhibited the nociceptive and vasodilatory effects of the endogenous peptide, CGRP (Durham 2004; Olesen et al. 2004). This non-peptide antagonist was developed by Boehringer Ingelheim GmbH, Germany, following a high-throughput screening (HTS) campaign around dipeptide derivatives (Bell 2014). Subsequently, studies on structure-activity relationship of these compounds led to discovery of an intermediate compound which was further optimized at its benzoxazolone ring to yield olcegepant (CHEMBL AlogP = 2.78) (Fig. 1) (Bell 2014). However, due to the difficulties encountered in developing an oral formulation, high molecular weight (MW) = 869 Da and polar surface area (PSA) = 181 Å^2, olcegepant studies were discontinued (Bell 2014; Kuzawinska et al. 2016).

2.1.1   Drug-Receptor Interaction Studies
The affinity of olcegepant for the GCRP receptor was found to be >100-fold higher for primate over non-primate receptors (Poyner et al. 2002; ter Haar et al. 2010).

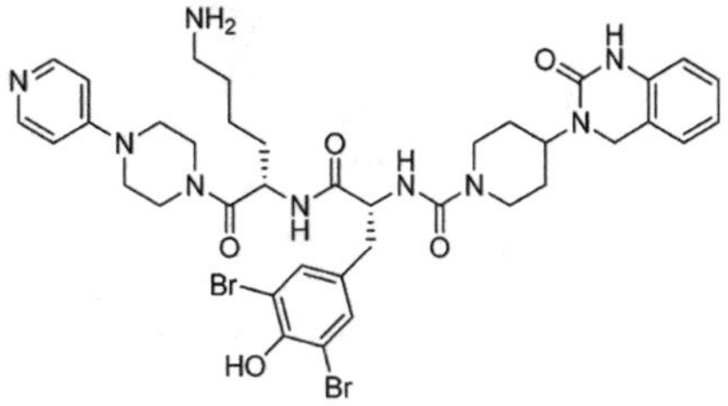

Fig. 1 Olcegepant. Source: Bell (2014)

Several studies revealed that olcegepant shows a specific affinity for the extracellular region of RAMP1, rather than the CLR receptor or the intracellular RCP subunit (Durham and Vause 2010). The crystallographic model of the extracellular domain of the CGRP receptor in complex with olcegepant (PDB 3N7S) showed that tryptophan at position 74 of the RAMP1 is the key residue for higher affinity of olcegepant for primate CGRP receptors (Fig. 2a) (Kandepedu et al. 2015; ter Haar et al. 2010); this important RAMP1 residue position (74) is highly variable outside of the primate family. The decrease in activity for rodent CGRP receptor is conferred by replacement of Trp 74 with Lys (Kandepedu et al. 2015). When olcegepant binds to the CGRP receptor, it stretches around 18 Å from a hydrogen bond donor site at Thr 122 of the CLR receptor into the deep hydrophobic pocket, consisting of the helix αC1 of the CLR receptor and helix αR2 of RAMP1 (ter Haar et al. 2010). Trp 74 (helix αR2) and Trp 84 (in the loop connecting helixes αR2 and αR3 of RAMP1) form the roof and the back surface of the binding pocket (ter Haar et al. 2010). Therefore, replacement of Trp 74 by lysine would reduce the ligand-protein hydrophobic area and sterically hinder the access to the hydrophobic pocket (ter Haar et al. 2010). Other RAMP1 amino acids that may play a small role in selectivity are represented by Ala 70, Asp 71, His 75, Phe 83, Trp 84, and Pro 85 (ter Haar et al. 2010).

One important movement encountered when the antagonist binds to the receptor is represented by the 70° rotation of the side chain of Trp 72 belonging to the CLR receptor. The rotation creates a "Trp shelf" where the piperidine ring of the antagonist settles (Fig. 2b) (ter Haar et al. 2010). The dibromotyrosyl group reaches deep into the pocket and binds by both hydrophobic and electrostatic interactions (ter Haar et al. 2010). Moreover, the carbonyl of the second amide bond forms a hydrogen bond with the NH of Trp 72 belonging to the CLR receptor (ter Haar

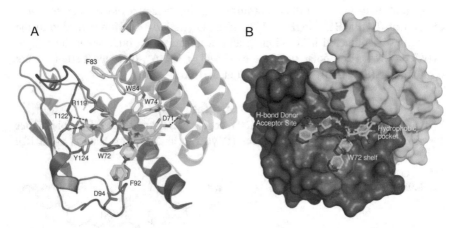

Fig. 2 Structures of olcegepant in complex with the CGRP receptor: (**a**) ribbon representation and (**b**) surface representation. The hydrogen bonds between olcegepant and the CLR receptor (pink) or the RAMP1 protein (cyan) are indicated with dashed lines. Interactions were taken from ter Haar et al. (2010)

et al. 2010). Olcegepant also forms a salt bridge between its Lys 6-amino terminus and the carboxyl of Asp 71, belonging to RAMP1 (ter Haar et al. 2010).

2.1.2 Antimigraine Effects of Olcegepant and Interaction with Other Receptors

Clinical studies revealed that olcegepant was effective for migraine only after intravenous administration and its effect was visible after 30 min and continued to improve over a few hours (Kuzawinska et al. 2016). The success of treating migraine with this antagonist achieved a rate of 60% using doses ranging between 0.25 and 10 mg (Kuzawinska et al. 2016). This success represented a major step in the development of CGRP antagonists for migraine. The most notable side effect encountered in patients during the phase I and II clinical trials was associated with paresthesias (Recober and Russo 2007; Kuzawinska et al. 2016). Furthermore, results from the trials demonstrated that no side effects regarding blood pressure or heart rate have been noticed after its administration (Durham and Vause 2010). Moreover, this drug had no constrictor effect on the superficial, radial, and cerebral temporary arteries and did not present cerebral blood flow changes (Benemei et al. 2017). The lack of the vasoconstrictor properties represents an advantage over the triptans, the most effective current abortive drugs in the treatment of migraine (Durham and Vause 2010). Interestingly, studies on anesthetized rat closed cranial models concluded that olcegepant appears to act outside of the blood-brain barrier (BBB), more specifically in the wall of meningeal arteries which do not have a BBB (Recober and Russo 2007; Benemei et al. 2017).

On the other hand, olcegepant is not efficient in treating migraine caused by factors other than CGRP. For example, Tvedskov et al. (2010) showed that this antagonist is not efficient in preventing migraine caused by the nitric oxide (NO) donor glyceryl trinitrate. The explanation for this is that NO does not cause headache attacks by releasing CGRP. These findings are in concordance with studies demonstrating that olcegepant exhibits an affinity for the binding site of the endogenous ligand at the CGRP receptor (Durham 2004). Instead, other studies revealed that olcegepant presents a relatively low antagonism for the amylin receptor 1 (AMY1), which is composed of the calcitonin (CT) receptor and RAMP1 (Hay and Walker 2017); this is important because CGRP can also bind to the amylin receptor (Hay et al. 2018). This is perhaps expected, since both CGRP and AMY1 receptors share RAMP1 as a common subunit. Moreover, Walker et al. (2017) examined the ability of olcegepant to block CGRP stimulation of intracellular signaling molecules relevant to pain (cAMP, p38, ERK 1/2, and CREB phosphorylation) in rat trigeminal ganglia neurons and transfected Cos7 cells. They showed that olcegepant antagonism of CGRP-stimulated cAMP accumulation in Cos7 cells transfected with CGRP and AMY1 receptors is approximately 130-fold more potent at the CGRP receptor than at the AMY receptor. Results also showed that for this pathway olcegepant is approximately 14-fold more potent in blocking CGRP rather than in blocking amylin at the AMY1 receptor in Cos7 cells. However, the selectivity of olcegepant for the AMY1 receptor depends on the pathway measured (Walker et al. 2017). Interestingly, a concentration of 1 μmol L^{-1} olcegepant is enough to block both receptors, but in this way the selectivity of the antagonist is lost (Hay and Walker 2017). However, the antagonism at the AMY1 receptor may be underestimated, and further studies are required to clarify its potential side effect (Walker et al. 2017).

2.2   Telcagepant (MK-0974)

MK-0974 (or telcagepant) was developed at Merck Research Laboratories under a program that aimed to discover the first orally active CGRP receptor antagonist. The benzodiazepinone moiety was identified as a potential lead after a HTS campaign (Bell 2014). Subsequent research led to the discovery of several compounds that underwent further optimization to obtain telcagepant (CHEMBL AlogP = 3.35), which presented an attractive combination of potency, selectivity, and oral bioavailability (Fig. 3) (Bell 2014).

2.2.1   Drug-Receptor Interaction Studies

Surprisingly, it has been shown that telcagepant (K_D = 1.9 nM) is not as potent as olcegepant (K_D = 45 pM) (ter Haar et al. 2010). However, while having a lower molecular weight, it has fewer but more productive interactions (ter Haar et al. 2010). The key substituent that provides high affinity for CGRP receptors is the 2,3-difluorophenyl substituent (Bell 2014). However, just like olcegepant, telcagepant is also RAMP-1 dependent and shows less affinity for non-primate species. Surprisingly, telcagepant showed 1,500 lower affinity compared to 100 lower affinity of olcegepant for these species (Kandepedu et al. 2015). One possible explanation for the huge difference between the antagonists for non-primate species could be the fact that telcagepant is smaller and not sterically hindered (Kandepedu et al. 2015). Therefore, replacement of tryptophan 94 in the human RAMP1 with lysine in rodents shows a higher impact on affinity for olcegepant (Kandepedu et al. 2015).

Despite its lower molecular weight (MW = 566 Da), telcagepant binds in a similar way to olcegepant by acting as a lever between Thr 122 of the CLR receptor and the hydrophobic pocket of RAMP1 and breaking nearby interactions which are important for peptide binding (Fig. 4) (Miller et al. 2010; ter Haar et al. 2010). Besides its interactions with Trp 74 and Trp 84, olcegepant presents an additional hydrogen bond between its azabenzimidazolone ring and the carbonyl belonging to

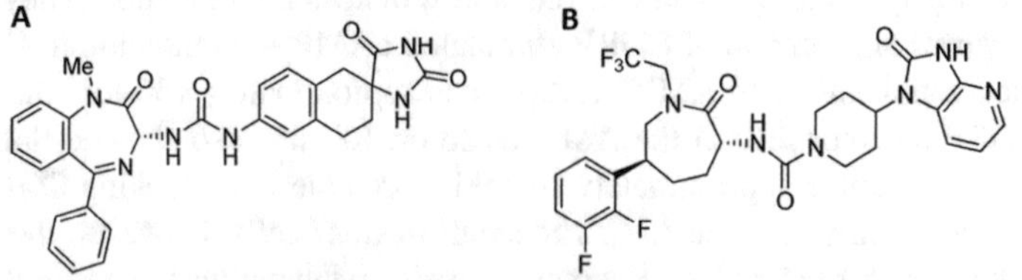

Fig. 3 Structures of the benzodiazepinone-based lead (**a**) and telcagepant (**b**). Source: Bell (2014)

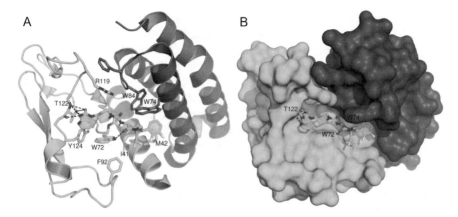

Fig. 4 Structures of telcagepant in complex with the CGRP receptor: (**a**) ribbon representation and (**b**) surface representation. The hydrogen bonds between telcagepant and the CLR receptor (pink) or the RAMP1 protein (cyan) are indicated with dashed lines. Interactions were taken from ter Haar et al. (2010)

Thr 122 of the CLR receptor (ter Haar et al. 2010). Moreover, the difluorophenyl group of telcagepant reaches deeper into the pocket than the dibromotyrosyl group of olcegepant, but it only relies on hydrophobic interactions with Met 42 of the CLR receptor (ter Haar et al. 2010). This residue, which is involved in binding of both antagonists, is more important for telcagepant affinity as shown by Miller et al. (2010). Another hydrophobic interaction of telcagepant arises between its trifluoromethyl group and Ile 41 of the CLR receptor.

2.2.2 Antimigraine Effects of Telcagepant and Interaction with Other Receptors

Like olcegepant, telcagepant is a highly selective antagonist for the GCRP receptor that can stop migraine pain and other migraine symptoms like nausea, photophobia, and phonophobia (Ho et al. 2008; Durham and Vause 2010). Telcagepant lacks vasoconstrictor properties which may allow a safe administration in patients suffering from migraine and cardiovascular disease, though further studies are required to assess the safety in this class of patients (Ho et al. 2008). Pharmacokinetic studies on telcagepant demonstrated that it has fairly good absorption, with plasma concentrations that decrease in a biphasic way and a half-life of about 6 h (Edvinsson and Linde 2010). Moreover, it showed relief from pain 30 min after administration, and a steady state was achieved in 3–4 days of multiple dosing (Edvinsson and Linde 2010). A phase II study compared the clinical effects of 25, 50, 100, 200, 300, 400, and 600 mg telcagepant with 10 mg of rizatriptan. Doses of 300, 400, and 600 mg were significant versus placebo, and 300 mg of telcagepant seemed to have outcomes as effective as rizatriptan (Edvinsson and Linde 2010). The study confirmed that telcagepant was effective in relieving pain at 2 h and provided sustained freedom from pain for up to 24 h (Edvinsson and Linde

2010). A phase III clinical study evaluated the efficacy and tolerability of telcagepant 150 and 300 mg in comparison with zolmitriptan 5 mg and showed that telcagepant (300 mg) was superior to telcagepant (150 mg) and placebo (Ho et al. 2008). Furthermore, telcagepant (300 mg) had a similar 2 h efficacy to zolmitriptan (5 mg) but showed fewer adverse effects than this compound (Ho et al. 2008). Another study also demonstrated that administration of 600 mg of telcagepant did not show any side effects on arterial blood pressure (Edvinsson and Linde 2010). However, data from a long-term safety trial reported that taking telcagepant (140 or 280 mg) twice daily for 12 weeks for migraine prevention led to raised concentration of liver transaminases (Edvinsson and Linde 2010). This suggests that the risk of liver toxicity may be dependent on dose and time (Kuzawinska et al. 2016). Consequently, concerns about hepatic toxicity terminated telcagepant development (Kuzawinska et al. 2016).

Regarding the interaction with other receptors, Walker et al. (2017) also revealed that telcagepant can act as an antagonist to the AMY_1 receptor in transfected Cos7 cells. Though, this antagonist was 35-fold more potent at the CGRP receptor compared to the AMY_1 receptor when CGRP-stimulated cAMP accumulation was measured (Walker et al. 2017). The antagonism of telcagepant when measuring CREB phosphorylation was not significantly different than that for cAMP accumulation. Telcagepant was approximately tenfold more potent at the GCRP receptor than at the AMY1 receptor. Surprisingly, compared to olcegepant, telcagepant selectivity seems to not depend on the pathway being measured (Walker et al. 2017).

2.3   BI 44370 TA

BI 44370 TA is another small-molecule antagonist at the GCRP receptor developed by Boehringer Ingelheim. It was discovered by focusing on reducing the molecular weight and polar surface area of olcegepant, as well as identifying an oral formulation (Fig. 5) (Bell 2014). In phase I trials, BI 44370 TA displayed good tolerability and minimal adverse effects (Diener et al. 2010). A phase II trial was conducted to assess the efficacy of three doses of BI 44370 TA, 50, 200, and 400 mg, for the treatment of acute migraine attacks (Diener et al. 2010). The three doses of BI 44370 TA were compared with eletriptan 40 mg and placebo. The study showed that the

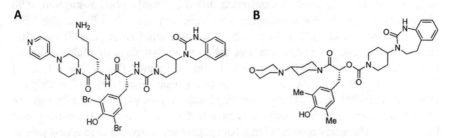

Fig. 5 Structures of olcegepant (**a**) and BI 44370 TA (**b**). Source: Bell (2014)

primary endpoint (pain-free at 2 h) was achieved by the subjects in the BI 44370 TA 400 mg and the eletriptan groups (Diener et al. 2010). The BI 44370 TA 400 mg group displayed similar endpoints as subjects treated with other CGRP antagonists (Diener et al. 2010). The BI 44370 TA 50 mg effect was similar to placebo, while the effect of the 200 mg BI 44370 TA was superior to placebo, but it failed to reach the primary endpoints (Diener et al. 2010). Moreover, the frequency of adverse effects was low in all the groups investigated, and no changes were found regarding ECG, pulse rate, and blood pressure (Diener et al. 2010). However, the development of the molecule was terminated for unknown reasons (Diener et al. 2015).

2.4   MK-2918

After the discovery of telcagepant, Merck Research Laboratories focused on developing another oral antagonist with a lower anticipated clinical dose than telcagepant (Paone et al. 2011). Therefore, they targeted improvements in potency and pharmacokinetic profile by increasing solubility and reducing plasma protein binding (Paone et al. 2011). An increased solubility, especially at acidic pH, was achieved by replacement of the caprolactam ring of telcagepant with imidazoazepane (Paone et al. 2011). In addition, the utilization of the azabenzoxazinone spiropiperidine structure decreased metabolism, and the tertiary methyl ether was found to be a good substituent for potency enhancement (Paone et al. 2011). Further optimization achieved the selection of MK-2918 (Fig. 6) (Paone et al. 2011). Studies showed that MK-2918 is more potent than telcagepant, but its bioavailability is only moderate in rats and low in dogs and rhesus monkeys (Paone et al. 2011; Bell 2014). However, after administration of MK-2918, substantial levels of an active metabolite (the alcohol derived from demethylation of the ether) were observed (Bell 2014). Therefore, it was expected that this metabolite contributes to the clinical efficacy, leading to a lower projected clinical dose (Paone et al. 2011; Bell 2014). Moreover, MK-2918 showed more than 6,000-fold selectivity in a panel of assays containing over 160 receptors, transporters, and enzymes and was selected as a preclinical candidate based on its profile although its current development status is still ambiguous (Paone et al. 2011; Bell 2014).

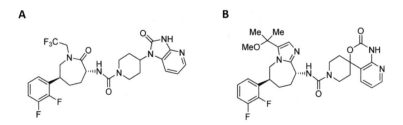

Fig. 6 Structures of telcagepant (**a**) and MK-2918 (**b**). Source: Bell (2014)

2.5   MK-3207

In parallel with the work that led to discovery of telcagepant and MK-2918 from the benzodiazepinone-based lead, the Merck group focused on developing another oral antagonist (Bell 2014). They followed an approach in which the spirohydantoin portion of the lead was retained and the benzodiazepinone group was replaced (Bell 2014). Further optimization of the spirohydantoin-based structure led to attractive intermediate lead structures which underwent several replacements to obtain MK-3207 (Fig. 7) (Bell 2014). The incorporation of a spirocyclopentyl-substituted piperazinone was the peak modification which allowed the optimal potency, selectivity, and pharmacokinetics of MK-3207 (CHEMBL MW = 557.6 Da, AlogP = 3.4) (Fig. 1) (Bell 2014). Detailed evaluation of MK-3207 revealed that it has a high affinity for the rhesus monkey GCRP receptor and a low affinity for the rat CGRP receptor (Bell 2014). Moreover, it was shown that MK-3207 is more potent than telcagepant both in vitro (>50-fold) and in vivo (>100-fold) (Bell 2014). Studies on binding using the tritiated analog [^{3}H]MK-3207 showed that the compound dissociated from the CGRP receptor more slowly ($t_{1/2}$ = 59 min) compared to telcagepant ($t_{1/2}$ = 1.3 min) (Bell 2014). Furthermore, the efficacy of MK-3207 in the treatment of migraine was evaluated in a randomized trial. Doses of 2.5, 5, 10, 20, and 50 mg of MK-3207 were chosen in the first part of the study. MK-3207 doses of 2.5 and 5 mg were shown to have insufficient efficacy, but only the 2.5 mg dose was discontinued from the study (Hewitt et al. 2011). In addition, due to the low efficiency of the other doses, a 200 mg dose was added in the second part of the trial (Hewitt et al. 2011). The study found that the pain-free rate after 2 h administration of 200 mg MK-3207 was superior to placebo and nominally significant for doses of 100 and 10 mg (Hewitt et al. 2011). Moreover, the authors concluded that the compound was well-tolerated and effective in acute treatment of migraine and the incidence of adverse effects (nausea, dizziness, sleepiness) did not appear to enhance with increasing dose (Hewitt et al. 2011). However, the studies on MK-3207 were soon terminated after it was found that MK-3207 caused elevated levels of transaminases (Bell 2014). This was a major discouraging effect in the global search for CGRP antagonists given that telcagepant had been discontinued due to the same adverse effect.

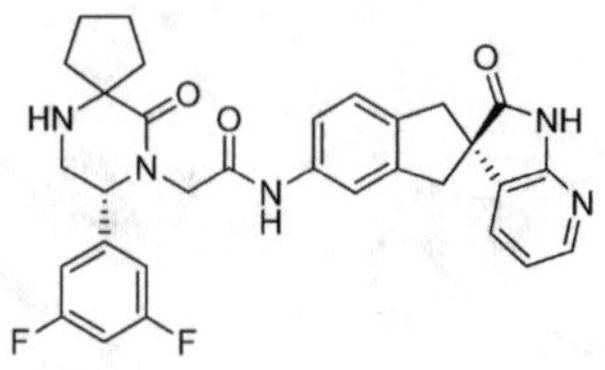

Fig. 7 Structure of MK-3207. Source: Bell (2014)

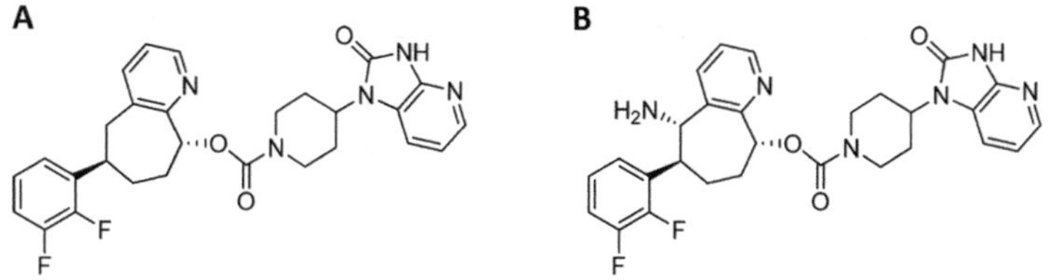

Fig. 8 Structures of BMS-846372 (**a**) and BMS-927711 (rimegepant) (**b**). Source: Bell (2014)

2.6   Rimegepant (BMS-927711/BHV3000)

Bristol-Myers Squibb also initiated a program that aimed to identify GCRP receptor antagonists for the treatment of acute migraine (Bell 2014). They identified a potent oral antagonist, BMS-846372, that contained a cyclohepta[b]pyridine core (Fig. 8a). The compound had a high resemblance to telcagepant and was an attractive clinical lead (Bell 2014). However, it proved to have a very low aqueous solubility (CHEMBL AlogP = 4.2) and created a significant challenge for further development (Luo et al. 2012). Several attempts like formation of salts of the core pyridine or phosphate-related prodrugs were made to address this problem, but they were unsuccessful (Luo et al. 2012). Therefore, they hypothesized that addition of an amino substituent to the cyclohepta[b]pyridine core will improve the solubility and will maintain the attractive properties of the lead compound (Luo et al. 2012; Bell 2014). This led directly to the discovery of BMS-927711 or rimegepant (CHEMBL MW = 534.6 Da, AlogP = 2.9) which presented substantial improvement in aqueous solubility and no serious challenges regarding development (Fig. 8b) (Bell 2014). Moreover, BMS-927711 was a more potent antagonist both in vitro and in vivo and displayed a good oral bioavailability in both rats and monkeys (Bell 2014). Currently, BMS-927711, rimegepant (recently named BHV-3000), is under development by Biohaven Pharmaceuticals. BMS-927711 efficacy was evaluated in a phase II trial at doses of 10, 25, 75, 150, 300, and 600 mg, with sumatriptan 100 mg and placebo as comparators (Marcus et al. 2014). The study showed that doses of 75, 150, and 300 mg were superior to placebo regarding being pain-free at 2 h after administration (Marcus et al. 2014). The 150 mg dose was the most effective in this case. However, for this endpoint, the 600 mg dose was not superior to placebo, and one reason for this may be because of the inherent variability of the patients in this group (Marcus et al. 2014). For the other endpoints, such as sustained pain-free to 24 h post-dose, doses ranging from 25 to 600 mg were all superior to placebo (Marcus et al. 2014). The incidence of adverse effects was low, and the most common effects were nausea, dizziness, and vomiting (Bell 2014; Marcus et al. 2014). To this end, the authors concluded that BMS-927711 is superior to placebo and has an excellent tolerability profile (Marcus et al. 2014). A phase III clinical trial to assess the efficacy of rimegepant versus placebo started in the middle of 2016 and is expected to come to completion at the end of March 2018 (ClinicalTrials.gov: NCT03237845). Another active phase III study with a focus on the tolerability and safety of rimegepant is estimated to conclude in April 2019 (ClinicalTrials.gov: NCT03266588).

2.7   BHV-3500

Another compound acquired by Biohaven Pharmaceuticals from Bristol-Myers Squibb is BHV-3500 for the prevention of episodic and chronic migraine (www. biohavenpharma.com). According to Biohaven, BHV-3500 is a highly soluble, potent, and selective drug at the human CGRP receptor. Preliminary preclinical evaluations on the marmoset model following oral delivery showed no significant cardiovascular safety or systemic toxicity concerns (www.biohavenpharma.com). Biohaven also reported that they are confident in BHV-3500's chemical properties, which allow multiple routes of delivery such as oral, nasal, inhalation, or subcutaneous administration (www.biohavenpharma.com). At the moment, Biohaven is progressing to submit an investigational new drug application (IND) to the FDA in the first half of 2018. The company also plans to conduct a phase I trial in the second half of 2018 to evaluate the pharmacokinetics, safety, and tolerability of BHV-3500 in healthy volunteers (www.biohavenpharma.com).

2.8   Ubrogepant (MK-1602)

Ubrogepant or MK-1602 (CHEMBL MW = 549.5 Da, AlogP = 2.9) is a novel small molecule drug that has been identified to act as a receptor antagonist to CGRP for the treatment of migraine (Voss et al. 2016). Initially, ubrogepant was developed by Merck Research Laboratories. However, in July 2015, Merck signed a licensing agreement with Allergan to cede exclusive worldwide rights to the new CGRP Migraine Development Program (www.allergen.com). Ubrogepant could be obtained after an amide bond formation between two intermediates, an amino lactam and a spiro acid (Fig. 9) (Yasuda et al. 2017). The preparation of these two intermediates represented a real synthetic challenge, and the discovery of new routes to these compounds became essential for the program to go further into clinical trials (Yasuda et al. 2017). For example, Yasuda et al. (2017) reported the asymmetric synthesis of the lactam intermediate by an enzyme mediated dynamic kinetic

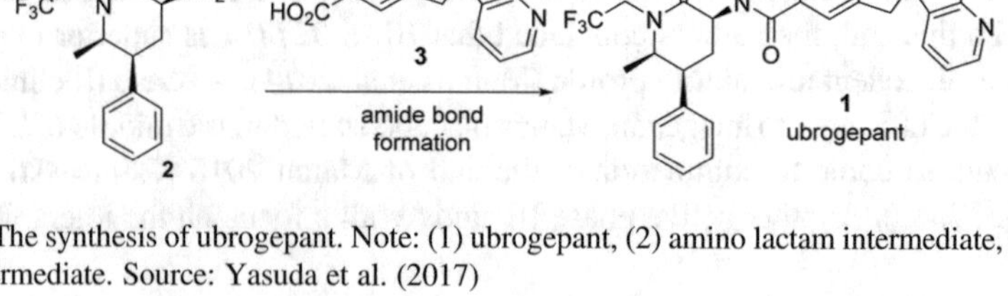

Fig. 9 The synthesis of ubrogepant. Note: (1) ubrogepant, (2) amino lactam intermediate, (3) spiro acid intermediate. Source: Yasuda et al. (2017)

transamination approach. They also described the asymmetric synthesis of the second intermediate using a novel doubly quaternized phase transfer catalyst spirocyclization (Yasuda et al. 2017).

Ubrogepant is a human p-glycoprotein substrate with moderate permeability that was shown to be absorbed quickly (T_{max} of 0.7–1.5 h) and to have a half-life of ~3 h for the α phase or ~5–7 h for the β phase (Voss et al. 2016). A phase IIb randomized, double-blind, placebo-controlled trial was conducted to assess the efficacy and tolerability of ubrogepant for the acute treatment of migraine (Voss et al. 2016). The doses of ubrogepant tested were 1, 10, 25, 50, and 100 mg. Ubrogepant 25, 50, and 100 mg were shown to be superior to placebo for the primary endpoint of 2 h pain-free, with the dose of 100 mg to be the most effective (Voss et al. 2016). However, for the 2 h headache response, only ubrogepant 100 mg showed a high response, but it was not significantly superior compared to placebo (Voss et al. 2016). Ubrogepant was well tolerated, and the side effects were similar to the ones encountered in the placebo group (Voss et al. 2016). The only two exceptions were for nausea and dizziness which were normally mild and self-limited. No elevated levels of liver enzymes were found in this present study (Voss et al. 2016).

ACHIEVE 1, a phase III study, that evaluated two doses of ubrogepant (50 and 100 mg) was completed at the end of 2017 (ClinicalTrials.gov: NCT02828020). Top-line results from Allergan showed that 19.2% of the patients in the 50 mg dose group and 21.2% of the patients in the 100 mg dose group experienced pain freedom at 2 h after the initial dose, compared to 11.8% patients in the placebo group. Moreover, the absence of the most bothersome migraine-associated symptoms including nausea, photophobia, and phonophobia were absent 2 h after the initial dose in 38.6% patients in the lower-dose group and 37.7% patients in the higher-dose group, versus 27.8% in the placebo set (www.allergan.com).

Another two phase III studies of ubrogepant began in the second half of 2016. ACHIEVE 2 evaluated the efficacy, safety, and tolerability of ubrogepant 25 and 50 mg for a single migraine attack (ClinicalTrials.gov: NCT02867709). It was completed in February 2018, and the results are expected in the first half of 2018. An extension phase III study to evaluate the long-term safety and tolerability of ubrogepant 50 and 100 mg is anticipated to end in October 2018 (ClinicalTrials.gov: NCT02873221). Allergan is expected to file for the new drug application (NDA) to the FDA in 2019 (www.allergan.com).

2.9 Atogepant (MK-8031)

Another CGRP inhibitor acquired by Allergan from Merck Research Laboratories following the licensing agreement is atogepant or MK-8031 (Fig. 10). Atogepant is the first oral CGRP antagonist being developed for migraine prophylaxis (www.allergan.com). Currently, atogepant is under a phase II/III study that will evaluate the efficacy, safety, and tolerability of once-daily dosing (10, 30, 60 mg) and twice-daily dosing (30, 60 mg) for the prevention of episodic migraine (ClinicalTrials.gov: NCT02848326). The study is expected to be completed by April 2018.

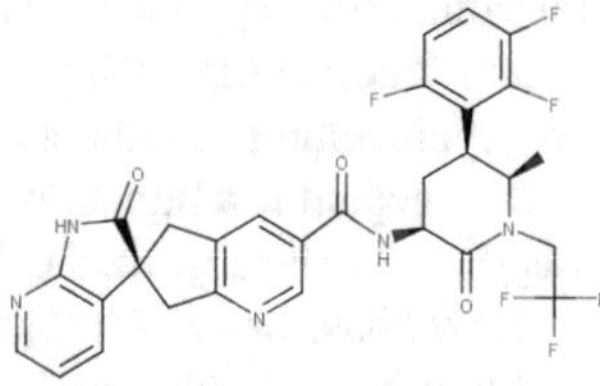

Fig. 10 Structure of
atogepant or MK-8031.
Source: PubChem

2.10   HTL0022562

HTL0022562 is a novel, potent, and highly selective CGRP antagonist designed by
Heptares Therapeutics (owned by the Sosei Group), using its structure-based drug
design platform. HTL0022562 was selected as the lead candidate following a
laborious selection process based on positive preclinical data under an alliance
with Teva Pharmaceutical Industries. A phase I trial in healthy volunteers is
expected to begin later in 2018 (www.heptares.com).

3   Anti-CGRP Ligand and Receptor Monoclonal Antibodies

A distinctive approach to inhibit GCRP and its receptor is represented by the
development of humanized monoclonal antibodies for the prophylactic treatment
of migraine (Kuzawinska et al. 2016). In comparison with gepants, monoclonal
antibodies (mAbs) present a slower onset of action, a prolonged half-life, an
enhanced target specificity, no abnormal liver side effects, a low risk of drug-drug
interactions, and the inability to cross the blood-brain barrier (Kuzawinska et al.
2016; Silberstein 2017). However, due to their large size, they can only be
administered parenterally (Silberstein 2017). Currently, there are four mAbs being
developed for migraine prevention.

3.1   Erenumab (AMG 334)

The only antibody that binds to the CGRP receptor is erenumab or AMG 334.
Erenumab was developed by Amgen Inc. However, in August 2015, Amgen started
a global collaboration with Novartis Pharmaceuticals which also includes the
erenumab program. AMG 334 is a fully human mAb of the IgG2 subtype that has
a half-life estimated at 21 days (Silberstein 2017).

The efficacy and safety of doses of AMG 334 of 7, 21, and 70 mg administered
subcutaneously for a period of 12 weeks to patients with episodic migraine were
evaluated in a phase II study (ClinicalTrials.gov: NCT01952574). The primary
endpoint was the mean change in monthly migraine days (MDM) from baseline to
week 12 of the study (Sun et al. 2016). A significant reduction of -3.4 days in the
primary endpoint from baseline at week 12 was encountered in the AMG 334 70 mg
group versus a reduction of -2.3 days in the placebo group ($p = 0.021$) (Sun et al.

2016). The lower doses did not significantly reduce migraine days. Adverse events such as fatigue, headache, and nasopharyngitis were mild to moderate and occurred in 50–54% of the patients in the AMG 334 groups compared with 54% in the placebo group (Sun et al. 2016). Nine patients also presented neutralizing antibodies (five in the AMG 334 7 mg group, three in the 21 mg group, and one in the 70 mg group) (Sun et al. 2016). This trial was followed by an open-label phase to assess the long-term safety and efficacy of AMG 334 for up to 5 years (Ashina et al. 2017). Results from the first completed year of this open-label follow-up evaluated changes in mean of migraine days (MMD) for patients who received AMG 334 70 mg every 4 weeks (Ashina et al. 2017). The researchers reported that an ongoing preventive effect exists, with a 5-day reduction in MMD in week 64 compared to the baseline of the initial study (Ashina et al. 2017). Moreover, at this week, 65, 42, and 26% of the patients achieved response rate of $\geq$50%, $\geq$75%, and 100%, respectively (Ashina et al. 2017).

Data from ARISE, a phase III study of erenumab 70 mg for episodic migraine, has recently become available (ClinicalTrials.gov: NCT02483585). The study showed that patients in the AMG 334 group experienced -2.9 days change in MMD versus -1.8 days for placebo group (Dodick et al. 2018). As well, a reduction in MMD of more than 50% was present in 29.5% in the placebo group and in 39.7% of the patients treated with AMG 334 (Dodick et al. 2018). Results from STRIVE (ClinicalTrials.gov: NCT02456740), a 6-month phase III trial, demonstrated a decrease in migraine days by 3.7 in the 140 mg dose of AMG 334, 3.2 in the 70 mg group, and 1.8 in the placebo group (Goadsby et al. 2017). Two other trials sponsored by Novartis Pharmaceutical are currently ongoing. Both studies are assessing AMG 334 for the treatment of episodic migraine. However, one evaluates AMG 334 in countries beyond the USA and the European Union by using a single-cohort, three-treatment arm (ClinicalTrials.gov: NCT03333109), while the other is evaluating AMG 334 in patients who have failed prophylactic migraine treatments (ClinicalTrials.gov Identifier: NCT03096834).

AMG 334 was also studied in a phase II trial for chronic migraine. The study demonstrated a reduction of -6.6 days in MMD for AMG 334 70 and 140 mg groups compared to -4.2 days for the placebo group (Tepper et al. 2017). The most frequent side effects were nausea, injection-site pain, and upper respiratory tract infection (Tepper et al. 2017).

All the findings from these phase II and phase III trials suggest that AMG 334 can be beneficial for the prevention of episodic migraine (Goadsby et al. 2017). While no cardiovascular safety concerns were identified in these trials, there is a theoretical cardiovascular risk associated with inhibition of the GCRP pathway (Ashina et al. 2017). Therefore, long-term studies are required to assess this matter and to determine the durability of AMG 334 effects (Ashina et al. 2017; Goadsby et al. 2017).

In July 2017, the AMG 334 license was accepted for review by the US Food and Drug Administration (FDA). AMG 334, trade name Aimovig, has recently been approved and is the first monoclonal antibody on the market that targets the CGRP receptor (www.amgen.com).

3.2 Fremanezumab (TEV-48125, LBR-101, PF-04427429, RN-307)

Fremanezumab, initially known as RN-307 and PF-04427429, is a fully humanized monoclonal antibody of the IgG2 isotype for the prevention of episodic and chronic migraine that was initially developed by Rinat Neuroscience, a company acquired by Pfizer in 2006 (Peroutka 2014; www.tevapharm.com). In 2012, after completing a phase I program for fremanezumab, Pfizer ceded the worldwide rights on the drug to Labrys Biologics. However, in June 2014, Labrys Biologics including the program that was in phase IIb clinical trials was acquired by Teva Pharmaceutical Industries Ltd. (www.tevapharm.com). In December 2017, Teva Pharmaceutical Industries announced that the FDA accepted a priority review for the license for TEV-48125 (www.tevapharm.com).

Unlike AMG 334 which binds to the CGRP receptor, TEV-48125 binds directly to CGRP and blocks its ability to interact with the receptor (Peroutka 2014). TEV-48125 presents a half-life of 39–48 days (Peroutka 2014). Results from five separate phase I trials showed that single intravenous infusions of TEV-48125 ranging between 0.2 and 2,000 mg and multiple infusions of doses up to 300 mg (once at 14 days) were well-tolerated (Peroutka 2014). However, a maximally tolerated dose was not identified (Peroutka 2014).

A phase IIb trial assessed the safety, efficacy, and tolerability of TEV-48125 225 mg and 675 mg administered subcutaneously in the preventive treatment of high-frequency episodic migraine (ClinicalTrials.gov Identifier: NCT02025556). The least square mean (LSM) change in the number of migraine days from baseline to weeks 9–12 decreased with -3.46 days in the placebo group compared to -6.27 days for the 225 mg dose group and -6.09 days for the 675 mg group (Bigal et al. 2015a, b). Adverse events took place in 46% of patients subjected to the lower-dose group, 59% patients in the higher-dose group, and 56% patients in the placebo group (Bigal et al. 2015a, b). The majority of reported adverse events were related to injection-site pain or erythema. No liver function abnormalities, cardiovascular changes, or immunological dysfunctions were found (Bigal et al. 2015a, b). However, two patients presented antibodies against TEV-48125 before and after the treatment. However, no increases in antibody titers were found, and no treatment-emergent antibody response was documented (Bigal et al. 2015a, b).

Other phase IIb trials assessed subcutaneous TEV-48125 675/225 mg and 900 mg for the preventive treatment of chronic migraine (ClinicalTrials.gov: NCT02021773). They randomly assigned patients to three 28-day treatment cycles of TEV-48125 (675 mg in the first cycle followed by 225 mg in the second and third cycle, 900 mg in all three cycles), or placebo (Bigal et al. 2015a, b). The mean decrease from baseline to week 9–12 in headache hours were significantly greater for the active groups (-60 h for the 675/225 mg group, -68 h in the 900 mg group) compared to placebo group (-37 h) (Bigal et al. 2015a, b). The LSM difference in the reduction of headache hours between placebo and active groups was also assessed. For the placebo and 675/225 mg groups was -23 h, while for placebo and 900 mg groups was -30 h. No serious treatment-related adverse effects were found (Bigal et al. 2015a, b).

Results from a 12-week phase III trial to assess the efficacy and safety of subcutaneously administered TEV-48125 225 and 675 mg regimens for the treatment of chronic migraine have recently become published (ClinicalTrials.gov: NCT02621931). Patients were randomly assigned to receive TEV-48125 quarterly (a single dose of 675 mg at baseline followed by placebo at weeks 4 and 8), monthly (675 mg dose at baseline, followed by 225 mg dose at weeks 4 and 8), or placebo (Silberstein et al. 2017). The mean number of headache days was reduced by 4.6 ± 0.3 days for the quarterly group, 4.6 ± 0.3 days for the monthly group, and 2.5 ± 0.3 days in the placebo group (Silberstein et al. 2017). The patients who received the monthly regimen had a 41% reduction in the average headache days per month, the quarterly regimen had a 38% reduction, and the placebo group had an 18% reduction (Silberstein et al. 2017). Other phase III studies of TEV-48125 for both episodic and chronic migraine preventive treatments are ongoing, as well as for cluster headache.

3.3 Galcanezumab (LY2951742)

Galcanezumab or LY2951742 is a humanized antibody of the IgG4 subtype with a half-life of 28 days that also targets CGRP for the prevention of migraine and cluster headache (Schuster and Rapoport 2016). It was initially developed by Eli Lilly and Co. In 2011, LY2951742 was licensed to Arteaus Therapeutics (Peroutka 2015). However, in 2014 Eli Lilly reacquired the rights to the experimental drug (Peroutka 2015).

In a phase II proof-of-concept study, LY2951742 150 mg given subcutaneously every 2 weeks for a duration of 12 weeks in patients suffering episodic migraine was assessed (ClinicalTrials.gov: NCT01625988). LY2951742 was found to be superior to placebo, with a mean change from baseline to week 12 in migraine days of -4.2 compared to -3.0 days (Dodick et al. 2014a, b). Moreover, the patients in the LY2951742 group also experienced high responder rates. The migraine headache was reduced by >50% in 70% of the patients and >75% in 49% of the patients, and there was complete elimination of the attacks in 32% of patients (Dodick et al. 2014a, b). Injection-site reactions and erythema were the most frequent adverse events that occurred in the active group. Viral infections and upper respiratory infections were the most common events in both groups (Dodick et al. 2014a, b).

Another phase II trial evaluated the effect of subcutaneous administration of galcanezumab 5, 50, 120, and 300 mg for the prevention of episodic migraine during a 3-month period (ClinicalTrials.gov: NCT02163993). The primary outcome for this study was represented by superiority, which was determined when the posterior probability of a greater boost for any active group compared with placebo measured by the mean change from baseline in the number of migraine headache days (MHD) in month 3 was $\geq 95\%$ (Bayesian analysis) (Skljarevski et al. 2018). The 120 mg dose of LY2951742 met the primary outcome and significantly reduced the mean of migraine days (99.6% posterior probability -4.8 MHD; 90% Bayesian credible intervals, -5.4 to -4.2 MHD) compared to placebo (-3.7 MHD, 90% BCI, -4.1

to −3.2 MHD) (Skljarevski et al. 2018). The overall change from baseline in the number of MHD was also significant for the 120 mg group (−4.3 MHD; 95% CI, −4.9 to −3.7 MHD; $P = 0.02$) and 300 mg group (−4.3 MHD; 95% CI, −4.9 to −3.7 MHD; $P = 0.02$) compared with the placebo group (−3.4 MHD; 95% CI, −3.8 to −2.9 MHD) (Skljarevski et al. 2018). The adverse effects that occurred in this study were consistent with other findings and included nausea, injection-site pain, dysmenorrhea, and upper respiratory tract infections (Skljarevski et al. 2018).

The efficacy and safety of several doses of galcanezumab were also studied in healthy volunteers in a phase I trial (ClinicalTrials.gov: NCT01337596). It was found that subcutaneous injections of LY2951742 were well tolerated in all single doses (1–600 mg) and consecutive doses (150 mg). LY2951742 was found to induce a durable, robust, and dose-dependent inhibition of the capsaicin model, a target engagement biomarker that leads to a CGRP-mediated increase in dermal blood flow (Monteith et al. 2017). The adverse effects were generally similar between active groups and placebo and included headache, nasopharyngitis, and contact dermatitis (Monteith et al. 2017). An increased level of alanine and aspartate aminotransferase were found in five patients in the active group, but researchers concluded that these were unrelated to LY2951742 (Monteith et al. 2017).

Currently, LY2951742 is being studied in several phase III trials. In EVOLVE-1 (ClinicalTrials.gov: NCT02614183)   and   EVOLVE-2   (ClinicalTrials.gov: NCT02614196), galcanezumab is evaluated for prevention of episodic migraines. The completion dates are October 2018 for EVOLVE-1 and April 2019 for EVOLVE-2. Galcanezumab is also undergoing evaluation in REGAIN, a phase III trial conducted for the prevention of chronic migraine with an expected completion date in July 2019 (ClinicalTrials.gov: NCT02614261). Two other phase III trials to assess the efficacy and safety of LY2951742 in episodic (ClinicalTrials.gov: NCT02397473) and chronic (ClinicalTrials.gov: NCT02438826) cluster headache are ongoing. The estimated completion dates are June 2018 and July 2019, respectively. The last phase III trial that is expected to complete in December 2018 aims to evaluate the long-term safety of LY2951742 in patients with episodic and chronic migraine (ClinicalTrials.gov: NCT02614287).

3.4   Eptinezumab (ALD403)

Eptinezumab or ALD403 is another humanized monoclonal antibody that binds selectively and potently to inhibit CGRP (Silberstein 2017). The plasma half-life after infusion is 31 days (Silberstein 2017). ALD403 is being developed by Alder Biopharmaceuticals for the preventive treatment of episodic and chronic migraine. Interestingly, ALD403 which is of the IgG1 subtype is produced using yeast, not mammalian cells (Kuzawinska et al. 2016; Silberstein 2017). Moreover, ALD403 is the only anti-CGRP antibody designed to be administered by quarterly infusion (www.alderbio.com).

A phase II trial investigated the safety and efficacy of ALD403 1,000 mg administered as a one intravenous infusion in patients suffering frequent episodic migraine (ClinicalTrials.gov: NCT01772524). The active group showed a reduction of 5.6 migraine headache days (MHD) in weeks 5–8 compared with baseline, while the placebo group presented a reduction of 4.6 days (Dodick et al. 2014a, b). Moreover, after the single infusion, the responder rate of 100% was found in 41% of the patients in the active group versus 17% in the placebo at weeks 9–12. The same responder rate was also found in 16% of the patients in the ALD403 group versus 0% in the placebo group at weeks 1–12 post-hoc analysis (Dodick et al. 2014a, b). Adverse events were experienced by 57% of the patients in ALD403 group and by 52% of the patients in the placebo group. Some of the most frequent events were upper respiratory tract infection, urinary tract infection, and back pain (Dodick et al. 2014a, b). Results from the anti-drug antibody assays suggested that 14% of the patients in the ALD403 group had the potential to form antibodies against ALD403 during the study. Reassuringly, the anti-drug titers were low, and no evident effects of immunogenicity on tested parameters were observed (Dodick et al. 2014a, b).

Results from a phase II trial assessing intravenously administered ALD403 in chronic migraine (ClinicalTrials.gov: NCT02275117) were presented at the fifth European Headache and Migraine Trust International Congress in September 2016 (Reichert 2017). The doses of eptinezumab evaluated were of 10, 30, 100, and 300 mg. The study showed that all doses were safe and well-tolerated (Silberstein 2017). The primary efficacy endpoint represented by the percentage of people that achieved a 75% reduction in migraine days (weeks 1–12) was met by 300 mg (33%) and 100 mg (31%) doses compared to placebo (21%) (Silberstein 2017). Moreover, ALD403 300, 100, and 30 mg showed a significant difference versus placebo for the mean change from baseline to weeks 1–12 in migraine days per month (Reichert 2017).

At the moment, there are three ongoing phase III trials that assess the efficacy and safety of ALD403 in migraine prevention. According to ClinicalTrials.gov, the PROMISE 1 study (ClinicalTrials.gov: NCT02559895) had an estimated completion date of December 2017. Top-line results were published on Alder Biopharmaceutical's webpage. PROMISE 1 evaluates ALD403 30, 100, and 300 mg administered intravenously once every 12 weeks in patients with frequent episodic migraine (www.alderbio.com). The results showed that patients in the 300 mg dose group had a reduction of 4.3 migraine days from baseline over weeks 1–12, while the patients in the 100 mg group experienced a reduction of 3.9 fewer migraine days per month compared to placebo (−3.2 days) (www.alderbio. com). Furthermore, in this period a responder rate of 75% reduction was achieved by 29.7% patients in the 300 mg group and 22.2% patients in the 100 mg group versus 16.2% patients in placebo set (www.alderbio.com). Over months 1–6, an average of one in five patients (20.6%) receiving ALD403 300 mg had a 100% responder rate with no migraine days. The safety profile was similar to placebo and consistent with other ALD403 studies (www.alderbio.com).

PROMISE 2, a second phase III trial that has an estimated primary completion date of June 2018, is studying ALD403 300 mg and 100 mg for the prevention of chronic migraine (ClinicalTrials.gov: NCT02974153). Top-line results were presented by the company in January 2018. The doses were administered as a single infusion once every 12 weeks. Patients in the study experienced an average of 16.1 migraine days per month at baseline (www.alderbio.com). Both doses met the primary endpoint. A reduction of 8.2 monthly migraine days from baseline over the 12 weeks versus 5.6 days for placebo was found. A responder rate of $\geq 75\%$ reduction from baseline was achieved by 33% of the patients versus 15% in the placebo group (www.alderbio.com).

An open-label study to assess the safety and tolerability of ALD403 in chronic migraine is also underway (ClinicalTrials.gov: NCT02985398). The study has an estimated primary end date of June 2018.

The results from these phase III trials will support the eptinezumab license submission to the FDA, which Alder Biopharmaceutical plans to file in the second half of 2018 (www.alderbio.com). If eptinezumab is approved, it will become the first CGRP antibody for migraine prevention that is administered as an infusion that allows for 100% of the dose to inhibit CGRP (www.alderbio.com).

4   Summary

Table 1 summarizes the history of the key compounds. The first gepant, olcegepant, was the first CGRP antagonist and showed that migraine could, in principle, be treated using CGRP antagonists. Telcagepant, the first orally active CGRP antagonist, was affected by concerns over liver toxicity under prolonged use, but some commentators think that these compounds were withdrawn too prematurely as the

Table 1 A selection of CGRP therapeutic entities, including those currently in clinical trial or due to enter clinical trial

Compound	Company	Progress
Small molecules		
Olcegepant	Boehringer	Phase II trials terminated; poor bioavailability
Telcagepant	Merck	Phase III trials terminated over concerns for liver tox
BI 44370 TA	Boehringer	Phase II trials terminated for unknown reasons
MK-2918	Merck	Preclinical candidate, development status unknown
MK-3207	Merck	Phase II trials terminated over concerns for liver tox
Rimegepant	Biohaven	Phase III ongoing
BHV-3500	Biohaven	Phase I on healthy volunteers to start in 2018
Ubrogepant	Allergan	Phase III
Atogepant	Allergan	Phase II ongoing
HTL0022562	Heptares	Phase I on healthy volunteers to start in 2018
Therapeutic antibodies		
Erenumab	Amgen	Approved
Fremanezumab	Teva	Phase III
Galcanezumab	Lilly	Phase III
Eptinezumab	Alder	Phase III

adverse effects were not detected with intermittent use (Holland and Goadsby 2018). Subsequent compounds, namely, ubrogepant and rimegepant, have progressed to phase III clinical trials, atogepant has progressed to phase II, and BHV-3500 and HTL0022562 are due to progress to phase I later this year. Encouragingly, liver toxicity has not been reported for these small molecule compounds (or for the antibodies), indicating that the liver toxicity observed for telcagepant and MK-3207 was related to the individual molecular structures rather than the strategy of targeting CGRP. Moreover, in contrast to the triptan alternatives, these compounds appear to lack vasoconstrictor properties (Bell 2014).

Regarding the antibody drugs, erenumab, which targets the CGRP receptor, has been approved while fremanezumab, eptinezumab, and galcanezumab, which target the CGRP peptide, are in phase III clinical trial (Table 1). The small molecules and the antibodies are complementary to each other, with both classes offering distinct therapeutic advantages. The advantage of antibodies is that they are highly specific for their target, and the use of humanized antibodies can minimize the risks of autoantibodies (which have been observed for the antibodies). Small molecules can be made highly specific by careful design, but off-target interactions are more likely to occur with small molecules. Small molecule CGRP antagonists are more likely to offer rapid relief at the onset of a migraine, but antibodies, with their longer duration of action following approximately monthly injections, potentially offer greater prophylaxis. The CGRP peptide is vasoactive and so is cardioprotective. While it is encouraging that a number of studies have noted no adverse cardiovascular effects, MaassenVanDenBrink et al. have suggested that the long-term removal of CGRP or blockage of the CGRP receptor, particularly by antibodies, may raise concerns for certain patient groups, including pregnant women and those with heart disease (MaassenVanDenBrink et al. 2016; Deen et al. 2017). On the other hand, unlike small molecules, antibodies are likely to be metabolized to their constitutive amino acids, and so adverse liver toxicity is highly unlikely.

References

Ashina M, Dodick D, Goadsby PJ, Reuter U, Silberstein S, Zhang F, Gage JR, Cheng S, Mikol DD et al (2017) Erenumab (AMG 334) in episodic migraine: interim analysis of an ongoing open-label study. Neurology 89(12):1237–1243

Barwell J, Wheatley M, Conner AC, Taddese B, Vohra S, Reynolds CA, Poyner DR (2013) The activation of the CGRP receptor. Biochem Soc Trans 41:180–184

Bell IM (2014) Calcitonin gene-related peptide receptor antagonists: new therapeutic agents for migraine. J Med Chem 57:7838–7858

Benemei S, Cortese F, Labastida-Ramirez A, Marchese F, Pellesi L, Romoli M, Vollesen AL, Lampl C, Ashina M (2017) Triptans and CGRP blockade – impact on the cranial vasculature. J Headache Pain 18:103

Bigal ME, Dodick DW, Rapoport AM, Silberstein SD, Ma Y, Yang R, Loupe PS, Burstein R, Newman LC et al (2015a) Safety, tolerability, and efficacy of TEV-48125 for preventive treatment of high-frequency episodic migraine: a multicentre, randomised, double-blind, placebo-controlled, phase 2b study. Lancet Neurol 14:1081–1090

Bigal ME, Edvinsson L, Rapoport AM, Lipton RB, Spierings ELH, Diener HC, Burstein R, Loupe PS, Ma Y et al (2015b) Safety, tolerability, and efficacy of TEV-48125 for preventive treatment of chronic migraine: a multicentre, randomised, double-blind, placebo-controlled, phase 2b study. Lancet Neurol 14:1091–1100

Deen M, Correnti E, Kamm K, Kelderman T, Papetti L, Rubio-Beltrán E, Vigneri S, Edvinsson L, MaassenVanDenBrink A (2017) Blocking CGRP in migraine patients – a review of pros and cons. J Headache Pain 18:96

Dickerson IM (2013) Role of CGRP-receptor component protein (RCP) in CLR/RAMP function. Curr Protein Pept Sci 14:407–415

Diener HC, Barbanti P, Dahlof C, Reuter U, Habeck J, Podhorna J (2010) BI 44370 TA, an oral CGRP antagonist for the treatment of acute migraine attacks: results from a phase II study. Cephalalgia 31:573–584

Diener HC, Charles A, Goadsby PJ, Holle D (2015) New therapeutic approaches for the prevention and treatment of migraine. Lancet Neurol 14:1010–1022

Dodick DW, Goadsby PJ, Silberstein SD, Lipton RB, Olesen J, Ashina M, Wilks K, Kudrow D, Kroll R et al (2014a) Safety and efficacy of ALD403, an antibody to calcitonin gene-related peptide, for the prevention of frequent episodic migraine: a randomised, double-blind, placebo-controlled, exploratory phase 2 trial. Lancet Neurol 13:1100–1107

Dodick DW, Goadsby PJ, Spierings ELH, Scherer JC, Sweeney SP, Grayzel DS (2014b) Safety and efficacy of LY2951742, a monoclonal antibody to calcitonin gene-related peptide, for the prevention of migraine: a phase 2, randomised, double-blind, placebo-controlled study. Lancet Neurol 13:885–892

Dodick DW, Ashina M, Brandes JL, Kudrow D, Lanteri-Minet M, Osipova V, Palmer K, Picard H, Mikol DD et al (2018) ARISE: a phase 3 randomized trial of erenumab for episodic migraine. Cephalalgia 38:1026–1037. https://doi.org/10.1177/0333102418759786

Durham PL (2004) CGRP receptor antagonists: a new choice for acute treatment of migraine? Curr Opin Investig Drugs 5:731–735

Durham PL, Vause CV (2010) CGRP receptor antagonists in the treatment of migraine. CNS Drugs 24:539–548

Edvinsson L, Linde M (2010) New drugs in migraine treatment and prophylaxis: telcagepant and topiramate. Lancet 376:645–655

Eftekhari S, Edvinsson L (2011) Calcitonin gene-related peptide (CGRP) and its receptor components in human and rat spinal trigeminal nucleus and spinal cord at C1-level. BMC Neurosci 12:112

Goadsby PJ, Reuter U, Hallstrom Y, Broessner G, Bonner JH, Zhang F, Sapra S, Picard H, Mikol DD et al (2017) A controlled trial of erenumab for episodic migraine. N Engl J Med 377:2123–2132

Hay DL, Walker CS (2017) CGRP and its receptors. Headache 57:625–636

Hay DL, Garelja ML, Poyner DR, Walker CS (2018) Update on the pharmacology of the calcitonin/CGRP family of peptides: IUPHAR review. Br J Pharmacol 175:3–17

Hewitt DJ, Aurora SK, Dodick DW, Goadsby PJ, Ge YJ, Bachman R, Taraborelli D, Fan X, Assaid C et al (2011) Randomized controlled trial of the CGRP receptor antagonist MK-3207 in the acute treatment of migraine. Cephalalgia 31:712–722

Ho TW, Ferrari MD, Dodick DW, Galet V, Kost J, Fan X, Leibensperger H, Froman S, Assaid C et al (2008) Efficacy and tolerability of MK-0974 (telcagepant), a new oral antagonist of calcitonin gene-related peptide receptor, compared with zolmitriptan for acute migraine: a randomised, placebo-controlled, parallel-treatment trial. Lancet 372:2115–2123

Holland PR, Goadsby PJ (2018) Targeted CGRP small molecule antagonists for acute migraine therapy. Neurotherapeutics 15:302–312

Kandepedu N, Abrunhosa-Thomas I, Troin Y (2015) Targeting a hotspot in calcitonin gene related peptide receptor; how to proceed. Indo Am J Pharm Res 5:205–208

Kuzawinska O, Lis K, Cessak G, Mirowska-Guzel D, Balkowiec-Iskra E (2016) Targeting of calcitonin gene-related peptide action as a new strategy for migraine treatment. Neurol Neurochir Pol 50:463–467

Liang YL, Khoshouei M, Deganutti G, Glukhova A, Koole C, Peat TS, Radjainia M, Plitzko JM, Baumeister W, Miller LJ, Hay DL, Christopoulos A, Reynolds CA, Wootten D, Sexton PM (2018) Cryo-EM structure of the active Gs-protein complexed human CGRP receptor. Nature 561:492–497

Luo G, Chen L, Conway CM, Denton R, Keavy D, Signor L, Kostich W, Lentz KA, Santone KS et al (2012) Discovery of (5S,6S,9R)-5-amino-6-(2,3-difluorophenyl)-6,7,8,9- tetrahydro-5H-cyclohepta[b]pyridin-9-yl 4-(2-oxo-2,3-dihydro-1Himidazo[4,5-b]pyridin-1-yl) piperidine-1-carboxylate (BMS-927711): an oral calcitonin gene-related peptide (CGRP) antagonist in clinical trials for treating migraine. J Med Chem 55:10644–10651

MaassenVanDenBrink A, Meijer J, Villalón CM, Ferrari MD (2016) Wiping out CGRP: potential cardiovascular risks. Trends Pharmacol Sci 37:779–788

Marcus R, Goadsby PJ, Dodick D, Stock D, Manos G, Fischer TZ (2014) BMS-927711 for the acute treatment of migraine: a double-blind, randomized, placebo controlled, dose-ranging trial. Cephalalgia 34:114–125

Miller PS, Barwell J, Poyner DR, Wigglesworth MJ, Garland SL, Donnelly D (2010) Non-peptidic antagonists of the CGRP receptor, BIBN4096BS and MK-0974, interact with the calcitonin receptor-like receptor via methionine-42 and RAMP1 via tryptophan-74. Biochem Biophys Res Commun 391:437–442

Monteith D, Collins EC, Vandermeulen C, van Hecken A, Raddad E, Scherer JC, Grayzer D, Schuetz TJ, de Hoon J (2017) Safety, tolerability, pharmacokinetics, and pharmacodynamics of the CGRP binding monoclonal antibody LY2951742 (galcanezumab) in healthy volunteers. Front Pharmacol 8:740

O'Connell JP, Kelly SM, Raleigh DP, Hubbard JA, Price NC, Dobson CM, Smith BJ (1993) On the role of the C-terminus of alpha-calcitonin-gene-related peptide (alpha CGRP). The structure of des-phenylalaninamide37-alpha CGRP and its interaction with the CGRP receptor. Biochem J 291:205–210

Olesen J, Diener HC, Husstedt IW, Goadsby PJ, Hall D, Meier U, Pollentier S, Lesko LM (2004) Calcitonin gene-related peptide receptor antagonist BIBN 4096 BS for the acute treatment of migraine. N Engl J Med 350:1104–1110

Pal K, Melcher K, Xu HE (2012) Structure and mechanism for recognition of peptide hormones by class B G-protein-coupled receptors. Acta Pharmacol Sin 33(3):300–311

Paone DV, Nguyen DN, Shaw AW, Burgey CS, Potteiger CM, Deng JZ, Mosser SD, Salvatore CA, Yu S et al (2011) Orally bioavailable imidazoazepanes as calcitonin gene-related peptide (CGRP) receptor antagonists: discovery of MK-2918. Bioorg Med Chem Lett 21:2683–2686

Peroutka SJ (2014) Calcitonin gene-related peptide targeted immunotherapy for migraine: progress and challenges in treating headache. BioDrugs 28:237–244

Peroutka SJ (2015) Clinical trials update 2014: year in review. Headache 55:149–157

Poyner DR, Sexton PM, Marshall I, Smith DM, Quirion R, Born W, Muff R, Fischer JA, Foord SM (2002) International Union of Pharmacology. XXXII. The mammalian calcitonin gene-related peptides, adrenomedullin, amylin, and calcitonin receptors. Pharmacol Rev 54:233–246

Recober A, Russo AF (2007) Olcegepant, a non-peptide CGRP1 antagonist for migraine treatment. IDrugs 10:566–574

Reichert JM (2017) Antibodies to watch in 2017. MAbs 9:167–181

Schuster NM, Rapoport AM (2016) New strategies for the treatment and prevention of primary headache disorders. Nat Rev Neurol 12:635–650

Silberstein SD (2017) Current management: migraine headache. CNS Spectr 22:4–12

Silberstein SD, Dodick DW, Bigal ME, Yeung PP, Goadsby PJ, Blankenbiller T, Grozinski-Wolff M, Yang R, Ma Y et al (2017) Fremanezumab for the preventive treatment of chronic migraine. N Engl J Med 377:2113–2122

Skljarevski V, Oakes TM, Zhang Q, Ferguson MB, Martinez J, Camporeale A, Johnson KW, Shan Q, Carter J et al (2018) Effect of different doses of Galcanezumab vs Placebo for episodic migraine prevention: a randomized clinical trial. JAMA Neurol 75:187–193

Sun H, Dodick DW, Silberstein S, Goadsby PJ, Reuter U, Ashina M, Saper J, Cady R, Chon Y et al (2016) Safety and efficacy of AMG 334 for prevention of episodic migraine: a randomized, double-blind, placebo-controlled, phase 2 trial. Lancet Neurol 15:382–390

Tepper S, Ashina M, Reuter U, Brandes JL, Dolezil D, Silberstein S, Winner P, Leonardi D, Mikol
    D et al (2017) Safety and efficacy of erenumab for preventive treatment of chronic migraine: a
    randomised, double-blind, placebo-controlled phase 2 trial. Lancet Neurol 16:425–434
ter Haar E, Koth CM, Abdul-Manan N, Swenson L, Coll JT, Lippke JA, Lepre CA, Garcia-Guzman
    M, Moore JM (2010) Crystal structure of the ectodomain complex of the CGRP receptor, a
    class-B GPCR, reveals the site of drug antagonism. Structure 18:1083–1093
Tvedskov JF, Tfelt-Hansen P, Petersen KA, Jensen LT, Olesen J (2010) CGRP receptor antagonist
    olcegepant (BIBN4096BS) does not prevent glyceryl trinitrate-induced migraine. Cephalalgia
    30:1346–1353
Voss T, Lipton RB, Dodick DW, Dupre N, Ge JY, Bachman R, Assaid C, Aurora SK, Michelson D
    (2016) A phase IIb randomized double-blind placebo-controlled trial of ubrogepant for the acute
    treatment of migraine. Cephalalgia 36:887–898
Walker CS, Raddant AC, Woolley MJ, Russo AF, Hay DL (2017) CGRP receptor antagonist
    activity of olcegepant depends on the signalling pathway measured. Cephalalgia:1–15. https://
    doi.org/10.1177/0333102417691762
Yasuda N, Cleator E, Kosjek B, Yin J, Xiang B, Chen F, Kuo SC, Belyk K, Mullens PR et al (2017)
    Practical asymmetric synthesis of a calcitonin gene-related peptide (CGRP) receptor antagonist
    ubrogepant. Org Process Res Dev 21:1851–1858